Alark Joshi

Art-inspired Techniques for Visualizing Time-varying Data

Alark Joshi

Art-inspired Techniques for Visualizing Time-varying Data

Novel Techniques for Effectively Visualizing Time-varying Data

VDM Verlag Dr. Müller

Imprint

Bibliographic information by the German National Library: The German National Library lists this publication at the German National Bibliography; detailed bibliographic information is available on the Internet at http://dnb.d-nb.de.

Cover image: www.purestockx.com

Publisher:
VDM Verlag Dr. Müller Aktiengesellschaft & Co. KG , Dudweiler Landstr. 125 a, 66123 Saarbrücken, Germany,
Phone +49 681 9100-698, Fax +49 681 9100-988,
Email: info@vdm-verlag.de

Zugl.: Baltimore, University of Maryland Baltimore County, Diss., 2007

Produced in USA and UK by:
Lightning Source Inc., La Vergne, Tennessee, USA
Lightning Source UK Ltd., Milton Keynes, UK
BookSurge LLC, 5341 Dorchester Road, Suite 16, North Charleston, SC 29418, USA

ISBN: 978-3-639-06769-9

To my loving family, without whose support this would not have been possible

ACKNOWLEDGMENTS

First and foremost I would most sincerely like to thank my advisor Dr. Penny Rheingans. She has been a great source of inspiration, motivation and support through all these years. She has taught me as much through her actions, as she has through her advice. I have gone often to her office thoroughly confused and stuck, only to come out with all the problems resolved and a clear mind.

I am very grateful to Dr. Marc Olano for his invaluable advice from time to time. He has always given sound, practical advice and is always willing to listen patiently and provide solutions to your problems. Dr. Marie desJardins has been an immense help throughout my stay at UMBC and through her course in Research Skills, taught me that discipline is a key ingredient in conducting research. I would like to thank Dr. Tim Oates for being a reader and providing immense help as I prepared this document. I would like to thank Dr. Lynn Sparling for being an extremely patient collaborator who has provided excellent feedback and continues to be a great source of all things physics. Last but not the least, I would like to thank Dr. Deborah Silver, who not only provided us with data but has also been very helpful in providing advice and suggestions from time to time.

The VANGOGH lab has always been a fun, crazy place to work in and it has only been so because of the wonderful friends I have made in the lab.

My stay at UMBC has been made particularly smooth by the wonderful friends in the department office. Marie Tyler helped me weave my way through the forms and rules from my very first day at UMBC and has been a great friend. Jane Gethman has been the person who has always had an answer for every question and has always been kind enough to spare a few moments to chat despite her busy schedule. Keara Fliggins and Vera Douglass have been extremely helpful throughout.

Last but definitely not the least, my amazingly supportive family has been the source of all my motivation and inspiration along the way. My brother's family, Ashutosh, Preeti, Rhea and Ryan have been very understanding and supportive. My sister-in-law, Nupur and her husband Kalyan have always provided moral support and helped us in every possible way through these years. I would like to most sincerely acknowledge the encouragement and advice that I have received from my in-laws, Dr. Leena Hiremath and Dr. Shirish Hiremath. Throughout my life, my parents, Nalini and Prakash Joshi have provided me with immense love, encouragement and inspiration. I can never thank them enough for always believing in me and giving me the very best in life. At this point, it seems incredibly hard to express, in words, my immense gratitude towards my wife, Minoti, who has provided the friendship, support, inspiration and love throughout all these years. I would like to sincerely acknowledge her sacrifice and thank her profusely for the same.

TABLE OF CONTENTS

LIST OF FIGURES

xvii

xxii

LIST OF TABLES

Chapter 1

INTRODUCTION

Visualization is the process of converting data into a visual representation. The aim of visualization is to provide application-domain experts with *insight* into their data. Visualizing medical data obtained using *Computed Tomography* (CT) and *Magnetic Resonance Imaging* (MRI) scans has become an integral part of the way that doctors diagnose and plan treatments. Using two- and three-dimensional visualizations, radiologists can visually identify anomalies and provide advice to specialists for further medical treatment. Similarly, scientists performing large-scale time-dependent experiments use visualization as an aid to understand their experiments. Visualization can help researchers to preempt or restart experiments with a different set of input parameters, if undesirable results are observed. In climatology, researchers study climate and large-scale weather patterns to reach an improved understanding of the underlying science. For example, an atmospheric physicist would prefer to visualize and track a hurricane or a typhoon as it develops over time, rather than looking at individual snapshots taken at regular intervals. The transformation of features within the phenomenon, as well as the overall direction of motion, can be invaluable to an expert.

The data generated from these experiments, scans, and simulations can be better understood by visualizing them to get provide insights for exploration or diagnosis purposes. *Time-varying data* are three-dimensional snapshots of a process or phenomena that has been captured at regular time intervals. Domains such as computational fluid dynamics, weather forecasting, and medical imaging (ultrasound/fMRI) generate time-varying data.

Time-varying data can be loosely divided into the following categories, based on the questions

1

that domain experts try to answer using visualization:

- Fields such as computational fluid dynamics have datasets that contain three-dimensional features that move and "morph" over time. The paths traversed by these three-dimensional features in time-varying datasets are of considerable interest to scientists. Visually or computationally tracking these features is a typical task undertaken by domain scientists. Scientists correlate the features and their motion in order to understand the path taken by one or more features of interest. The task of visually tracking features is further complicated by occluding features and their change in shape over time. We call this type of data *positionally variant* data.

- In some domains, the data is time-varying only in the form of the attribute that is varying over time (e.g., global rainfall readings for a decade). In this case, common tasks include identifying anomalies and observing trends (incremental or decremental) that need further investigation. Weather and climate researchers investigate such trends and anomalies on a regular basis to understand the overall change in the values over a time interval. One such example would be visualizing the variance in temperature values across the North American region over the last century. These kinds of datasets, with varying values of a quantity such as temperature or rainfall, are referred to as *value variant* data.

Visualization of time-varying data is a challenging problem, due to the complex task of assimilating the undergoing changes. A researcher may be interested in multiple regions or features of interest that are all changing and moving over time. Visually tracking and understanding the changes in the data can be impossible in certain cases and could potentially overload a user's cognitive abilities. As a result, change blindness may lead to incorrect inferences.

The naïve approach to visualizing time-varying data is to generate a snapshot for each individual time step using standard visualization techniques. This process of generating and observing snapshots relies heavily on the user's ability to identify and track regions of interest over time. However, the number of snapshots generated can be quite high (100-5000 in some domains), requiring considerable effort for the user to visually track features. Figure 1.1 shows one such exam-

FIG. 1.1. This set of snapshots shows successive time steps of a volume. As is evident from looking at the snapshots, it is very hard to correlate and track a particular feature over time.

ple with just five snapshots showing the motion of three features in a time-varying dataset. As can be seen, the direction of motion of the three features is not easily understood by just looking at the snapshots.

Some practitioners use animations to visualize and get insight into their time-varying data. However, animations also have limitations, such as inattentional blindness, that may affect the visual tracking capabilities of a practitioner (Tversky *et al.* 2002).

1.1 Illustrations

The human visual system is a complex and intricate system. Illustrations have been used extensively to convey information that is not easily conveyed by photographs (Hodges 1989). Illustrators are well aware of our visual and cognitive abilities. They purposefully produce illustrations that convey important information to the viewer by adding cues that will attract attention of the viewer.

1.1.1 Conveying motion through illustration

Snapshots are of limited use, because they produce a pictorial representation of the *current* timestep without any context regarding the *previous* or *subsequent* timesteps. The user is required to mentally complete the motion by looking at multiple snapshots. In the case of viewing an animation, research in psychology has proven that human beings can only remember information from five to ten frames before the current frame. An animation of time-varying data snapshots that contains over 100 frames will be of limited use and its effectiveness will heavily depend on the

4

skills of the expert examining the animation.

Because application domain experts are primarily focused on being able to visually track features as they move over time, neither of these techniques are particularly helpful for their tasks. Occlusions by other features and limited memory (in the case of visualizing more than a few timesteps of a huge dataset) can severely limit a user's capability to track a feature of interest over time.

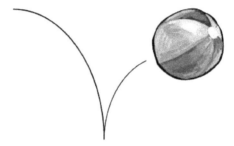

FIG. 1.2. This illustration shows a ball that has bounced in the past. The path traversed by the ball is conveyed using simple lines in a style used by illustrators to convey the motion of a cartoon character.

Illustrations communicate concepts to a reader in an easy-to-understand manner. Particularly, many children's book illustrators convey concepts using very few lines and simple, meaningful shapes. Figure 1.2, for example, shows an illustration similar to ones we may see in a children's book. It shows an image of a ball augmented by a line tracing an approximate path that the ball has traversed to get to its current location. It conveys past positions of the ball, shows where it originated from, and most importantly gives a dynamic quality to the static image of the ball. Many such examples show that illustrations - and comics in particular - are very effective at conveying motion or change in position over time in a subtle yet effective manner.

Leonardo Da Vinci used such techniques to convey motion over time. He used such techniques frequently in his scientific illustrations to convey change over time. He describes Figure 1.3 as follows:

FIG. 1.3. This figure, by Leonardo Da Vinci, contains four illustrations depicting the motion of water as it flows around an obstruction. Da Vinci draws similarities to the properties of hair as he discusses the flow of water around and over obstacles in its path (MacCurdy 1954).

"Observe the motion of the surface of the water, how it resembles that of hair, which has two motions - one depends on the weight of the hair, the other on the direction of the curls; thus the water forms whirling eddies, one part following the impetus of the chief current, and the other following the incidental motion and return flow." He used illustration principles to effectively convey concepts such as eddies and motion of water around an obstacle.

Illustrators have used numerous techniques to depict change over time in a single image (Mc-Cloud 1994). We are all familiar with comics where the illustrator uses techniques such as motion blur, faded representations of older positions, silhouettes of older positions, and similar technique to convey motion and action in scenes. Figure 1.4 shows one such example, where the downward motion of the axe is conveyed using trailing silhouettes.

FIG. 1.4. The downward motion of the axe is conveyed using trailing silhouettes. Illustration provided courtesy of Kunio Kondo (Kawagishi *et al.* 2003).

Illustrators such as Scott McCloud have discussed techniques for conveying motion from the perspective of an artist (McCloud 1994). Particularly speedlines and motion blur. McCloud uses Figure 1.5 to show how the illustrator conveys the flying motion of the man by adding lines that show the direction of the man's motion. The image, though static, successfully conveys motion and provides information regarding the original location of the man.

Figure 1.6 shows four different ways of conveying the rightward motion of a man using various techniques. The top left image conveys motion without any past information, whereas the top right and bottom left show some information regarding past positions of the man in the form of silhouettes or faded representations of the man. The bottom right image depicts a technique inspired by Japanese art, in which the motion is conveyed by implying that the scene around the

FIG. 1.5. This illustration depicts the flying motion of a man. The motion lines convey the direction of motion of the man and provide context to the current location of the man with respect to older locations in a single image. Image credits: McCloud (1994)

person is moving more than the person himself/herself. This can be seen in many Japanese anime cartoons, such as when a character is racing on a motorbike, but the scene around him is shown to be moving more than the motorbike.

1.1.2 Effective visualization of positionally variant data

In this work, we adapt such illustration principles to visualize time-varying data. For *positionally variant* data, we adapt techniques from the field of comics to convey positional change in a single image to the viewer. Four specific techniques were identified; namely speedlines, flow ribbons, opacity-based modulation and strobe silhouettes to annotate static images with motion

8

FIG. 1.6. This figure depicts four different techniques discussed by Scott McCloud to convey motion. The top-left image does not show any information about past positions, the top-right and bottom-left images clearly show some information regarding past positions and the bottom-right image depicts a style where the scene around the subject is shown to be moving to convey motion to the viewer. Image credits: McCloud (1994)

cues. From the conducted user study, it was found that illustration-inspired techniques helped users complete tasks such as visual tracking more accurately and confidently than with traditional visualization techniques. Participants consistently preferred illustration-inspired techniques over standard snapshot-based visualizations or animations.

The strengths and weaknesses of new visualization techniques are more evident when applied to an application domain. I applied our illustration-inspired techniques to visualize hurricane data. An expert evaluation was conducted to obtain the feedback of atmospheric physicists. They found our techniques to be useful for investigating hurricane structures as well as phenomena that can lead to intensification or dissipation in a hurricane.

1.1.3 Effective visualization of value variant data

Value variant data consists of an attribute changing values over time within the data. To visualize the change in values over a time interval or identify trends in a dataset, snapshots and animations are of limited use. I introduce pointillism-based techniques, inspired by the impressionist painter Seurat. Seurat, a keen student of color theory, applied the color-mixing theory that was developed by Chevreul and Rood (Rood 1879) to generate expressive paintings.

I developed techniques to paint a canvas using pointillistic strokes to represent the change undergone by a particular attribute over the specified time interval. A formal user evaluation of the techniques clearly indicated that the participants preferred pointillism-based techniques over static screenshots of time-varying data. The user study conducted revealed that participants were better at visually identifying variability as well as trends in the data using pointillism-based techniques than they were using standard visualization techniques.

1.2 Contributions

The research contributions to the field of visualization are in the form of novel visualization techniques inspired by art to visualize time-varying data. These techniques can be used to provide temporal context to the viewer by conveying motion as well as change over time more effectively.

Specifically, the contributions are as follows:

- Flow illustration techniques facilitate tracking of features over time. My work allows a viewer to better visualize the path traversed by a feature in a single image.

- The semantic simplification techniques maintain path features that are crucial to the process and produce simplified representations to mimic an illustrator's representation. Task-based simplified representations can be generated using the semantic simplification techniques.

- Pointillism-based visualization provides novel ways to visualize change in values attained by attributes in time-varying data. Using the developed techniques, trends in time-varying data can be clearly seen and used for understanding and conveying information about the properties of the attribute as it changes over time.

The art-inspired techniques have led to the following contributions in the application domains:

- In the field of fluid flow visualization, our flow illustration techniques provide improved visual tracking of features of interest over a large number of timesteps. My user evaluation provided convincing results that our techniques work better than standard snapshots or animations.

- Atmospheric physicists found our art-inspired techniques to be useful in the process of investigating a hurricane as it evolves over time. The internal structure of the hurricane, the path of the hurricane, and changes in attribute values such as humidity and temperature in the hurricane can now be better visualized using our techniques. My expert user evaluation provided crucial feedback that our techniques were preferred in most cases over visualizations generated by existing tools. My visualizations provided insight and helped them answer crucial questions and helped give them an increased understanding into the science behind the development of a hurricane.

These techniques and outcomes are a significant contribution to the field of visualization as well as the application domains.

1.3   Book Roadmap

The remainder of this book is organized as follows. In Chapter 2, I describe my work on illustration-inspired visualization of time-varying data for positionally variant data. In Chapter 3, I present results of a formal user study of those techniques and validate their effectiveness. In Chapter 4, I discuss various representations of the paths introduced in Chapter 3 and identify *smart* representations of those paths, which preserve important features based on the context.

In Chapter 5, I discuss a study describing the application of our techniques in the field of hurricane visualization. My techniques were validated by the expert user evaluation in which I obtained feedback from researchers in the fields of atmospheric physics, hurricane modeling, and simulations.

In Chapter 6, I describe our pointillism-inspired techniques for visualizing value variant data. I discuss our results in context of the hurricane domain as well as other $2D+t$ domains, such as global rainfall, US presidential election results for counties across the nation, and infant mortality data. In Chapter 7, I discuss the results of the formal evaluation of these pointillism-based techniques. We conclude in Chapter 8 with a discussion regarding the contributions and directions in which this work could be extended in the future.

Chapter 2

RELEVANT WORK

With an ever-increasing interest in illustration- and art-inspired computer graphics, the application of abstraction principles to the field of visualization is a natural extension to the concepts explored in non-photorealistic computer graphics. Non-Photorealistic Rendering (NPR) applies techniques from the field of art to generate images that have an illustrative or a painterly look.

NPR techniques have been applied to visualize data obtained from scientific experiments, medical scans, and other applications. These techniques have shown great promise in allowing users to explore medical CT scanned data and obtain information not readily available from traditional visualization techniques such as *volume rendering*. Many researchers have used such NPR techniques for visualizing medical and scientific datasets (Treavett and Chen 2000), (Rheingans and Ebert 2001), (Lu *et al.* 2002), (Lum and Ma 2002), (Csèbfalvi *et al.* 2001), (Hadwiger *et al.* 2003). These novel techniques have generated considerable interest in the field of medical volume visualization due to their ease of use and intuitive approach. Figure 2.1 shows a few examples of illustration-inspired techniques being applied to medical volumetric data. The top left image depicts a visualization that automatically accentuates features in the data. The boundaries and silhouettes in the data are emphasized in this case. The honeycomb structure in the liver can be clearly seen in this representation. The emphasized boundaries of organs helps to disambiguate their spatial location. The top right image is designed to draw the viewer's attention to a region of interest focused around the right kidney. The details in the right kidney are clearly seen as compared to the left kidney. The bottom left image shows an expressive visualization obtained by using the stippling style from the field of illustrations. The stippling technique provides an ex-

11

cellent representation that highlights features of interest and de-emphasizes unimportant regions. The bottom right image depicts a representation where features are highlighted by identifying and enhancing their curvature. Regions of high curvature are thickened in addition to the illustrative shading, generating an effective illustrative visualization.

Illustration-inspired techniques are very powerful at conveying information to a viewer in an informative yet pleasing manner.

2.1 Visualization of positionally variant data

Most research in the field of visualizing time-varying datasets has been focused around the needs of the application domain (computational fluid dynamics, medical imaging, weather fore-casting) from which the data was obtained. Visually examining and exploring the data often provides scientists with insight. Initial approaches dealt with visualizing each timestep of these huge datasets by generating a visualization using volume rendering or isosurface extraction techniques to represent the time-varying data (Weigle and Banks 1998), (Bajaj *et al.* 1999), (Chiang 2003).

Computational fluid dynamics (CFD) simulations were one of the first application domains that generated large amounts of time-varying data: that is, data in which three-dimensional features are moving and morphing over time. Samtaney *et al.* (1994) were the first to identify and apply techniques from the field of computer vision to deal with the problem of tracking these features in three-dimensional time-varying data. Various kinds of feature interactions such as merging of two features, splitting of one feature into two features, birth of a feature and death of a feature were characterized. Feature tracking was further improved by Silver *et al.* (1996), (1997). Figure 2.2 shows one of their results for feature tracking in the turbulent vortex dataset (Fernandez and Silver 1998) . Post *et al.* (1995) discussed "icon-based" techniques to visualize features in time varying data. Frameworks such as ViSTA FlowLib were developed to visualize flows in virtual environments (Schirski *et al.* 2003).

In addition to feature tracking and visualization, research in time-varying data visualization focuses on using compression (Shen *et al.* 1999), (Fout *et al.* 2005), (Westermann 1995) and optimized data structures and algorithms (Sutton and Hansen 1999), (Wilhelms and Gelder 1990),

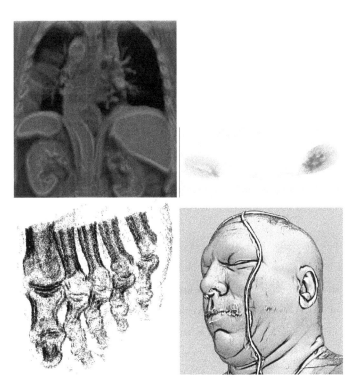

FIG. 2.1. This figure shows some innovative visualization techniques that lead to expressive visualization of medical volumetric data. The top left image shows a boundary and silhouette enhanced representation of an abdominal CT dataset, in which the internal honeycomb structure of the liver is visible along with the highlighted boundaries of the organs. The top right image draws the viewer's attention to an enhanced region of interest in the right kidney (Rheingans and Ebert 2001). The bottom left image shows a stipple-style rendering of a foot with the bones and the skin around it being emphasized. In particular, the joints are accentuated, drawing the viewer's attention in a manner similar to an illustrator (Lu *et al.* 2002). The bottom right image shows a representation of the head of the Visible Human dataset where the features on the head are highlighted using curvature-based enhancements (Kindlmann *et al.* 2003).

14

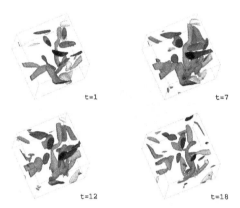

FIG. 2.2. This figure shows features from four timesteps of the turbulent vortex data that was used for feature extraction and tracking (Silver and Wang 1996).

to manage data efficiently. Recent advances in graphics hardware have also been leveraged to interactively visualize time-varying data (Lum *et al.* 2001). Cluster-based visualization over a distributed network has been developed to visualize huge time-varying datasets and to utilize the capabilities of the CPU and the GPU of each component in the system (Strengert *et al.* 2005), (Ma and Camp 2000). Research in the field of visualizing distributed datasets tackled the problem of visualizing prohibitively large datasets (Silver and Kusurkar 2000).

2.2   Illustrative visualization of positionally variant data

Despite their success, traditional visualization techniques are limited at conveying information to a user in applications with large amounts of data. As the datasets get bigger, using smart techniques to convey the information contained within the data becomes more important. Illustrative visualization techniques generate visualizations that can provide insight to the users into their data. They can highlight features of interest and draw the viewer's attention to those interesting features. Such techniques are invaluable for visualizing large datasets; without them the size of the dataset could cause an expert to inadvertently miss crucial features using standard visualization techniques.

Stompel *et al.* (2002) have applied non-photorealistic rendering techniques to time varying data visualization. They used techniques such as gradient, silhouette, and depth enhancement to provide more spatial and temporal cues. Figure 2.3 shows a comparison of a standard volume rendering of a CFD simulation on the left and an enhanced version on the right that accentuates internal structures using gradient, silhouette, and depth-based enhancements. Chronovolumes (Woodring and Shen 2003) were a novel technique developed to visualize time-varying data by using principles from art and photography. The ability to depict change as well as the overall motion was explored using these photography-based techniques. Svakhine *et al.* (2005) extended volume illustration techniques (Rheingans and Ebert 2001) for time-varying data and applied Schlieren photography and shadowgraphy from the field of photography to convey change over time. Figure 2.4 shows an example of an illustrative visualization of the temperature variation in a convection dataset. A specific temperature isocontour is shown, with other temperature regions shown as sketches around it.

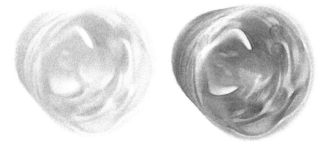

FIG. 2.3. This figure shows volume rendered images of a CFD simulation. The left image shows a standard volume rendering image and the right image shows a temporal domain enhanced version (Lum *et al.* 2001).

Other than the photography-inspired work by Svakhine *et al.* (2005), there is no work that specifically focuses on providing temporal context to the viewer regarding previous timesteps, or providing information regarding the timesteps in a certain interval around the current timestep be-

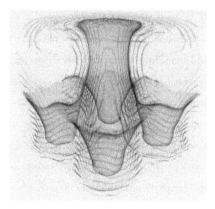

FIG. 2.4. An illustrative visualization of the temperature variation in a convection dataset. The image shows a specific temperature region rendered as a thin boundary volume and other temperature regions shown as sketches around it. (Svakhine *et al.* 2005)

ing visualized. Even in the case of photography-inspired visualizations, new visualizations are generated to provide temporal context (Svakhine *et al.* 2005). An expert evaluation of such photography-inspired techniques will help identify their true value for scientific visualization.

Conveying temporal context using illustration-inspired cues in a non-disruptive manner can be invaluable. A visual representation of the current timestep augmented by illustration-inspired techniques can not only provide context regarding a particular feature's movement, but can also help the viewer visually track features in feature-rich time-varying data. There is a need for investigating the illustration literature to identify effective, easily comprehensible techniques that can be applied to generate effective visualizations of time-varying data. An evaluation of the strengths and weaknesses of such techniques can provide insight into the effectiveness of such techniques and provide guidelines regarding the types of techniques to be used for various kinds of time-varying datasets.

2.2.1 Illustrative visualization of paths

My work deals with augmenting time-varying data visualizations with cues inspired from illustrations and comics. One of the techniques is called *speedlines* where the direction of motion of a three-dimensional feature is conveyed from an illustrator's point of view. To mimic an illustrator's depiction of the direction of motion, we generate a *simplified* representation of the actual path taken by a feature of interest. There has been a significant amount of work dealing with the problem of automatically generating simplified representations of large data. In our case, the data is in the form of the path traversed by a feature. In this subsection, we discuss relevant work in the field of path simplification.

Simplification can be defined as the process of generating a low-detail representation of a high-detail dataset; where the salient characteristics from the high-detail representation are preserved in the low-detail representation. Early work in the field of level-of-detail techniques for computer graphics was done by Clark (1976), Funkhouser and Sequin (1993) and Hoppe (1996). There has been some research in the use of visual perception-based knowledge to generate simplified representations of complex models (Luebke and Hallen 2001), (Cohen *et al.* 1998).

Cartographers often use simplification to generate simplified maps of regions (Robinson and Sale 1969). They frequently vary the simplification parameters of a map based on the function of the map. Research in the field of generating easy-to-use, task-based representations of maps has provided guidelines for the needs and benefits of using such simplified representations (Tversky 1992). The Douglas-Peucker algorithm was one of the first algorithms to generate a simplified representation of a path or loop given a large number of points as input (Douglas and Peucker 1973).

In the field of computer graphics, there has been some seminal work on the problem of generating simplified representations of curves, polygons, paths and maps. Chan and Chin (1996) discussed early ideas to produce a simplified representation of polygonal curves. Agarwal *et al.* (2000) developed efficient algorithms for simplifying curves and paths. Agrawala and Stolte (2001) were the first to propose a unique solution to the problem of generating comprehensive and easy-to-use route maps. Figure 2.5 shows one such visualization of driving directions generated by their sys-

18

tem; this visualization incorporates layout, design and comprehensibility constraints. Researchers have also developed graphics hardware-based algorithms to efficiently perform view-dependent generalization of maps in an interactive map visualization system (Mustafa *et al.* 2001).

FIG. 2.5. This is an image of a novel technique to visually represent driving directions data. Easy-to-use driving directions with context information were generated and simple cognitive cues were provided to the user. The user study conducted confirmed the effectiveness of their clutter-free visualization of driving directions (Agrawala and Stolte 2001).

The major limitation of all these techniques for my purposes is their inability to generate task-based visualizations for paths. Simplifying path data to generate a simplified representation which contains fewer points is useful for storage and rendering purposes, but it is not as useful for visualization purposes if the simplification does not take into account features along the path. Inadvertently, a path simplification algorithm could eliminate a crucial feature, leading to an incorrect interpretation of data.

There is clearly a need to identify regions of motion that are interesting from an application point of view and to ensure that they are preserved even in simplified representations of the data.

2.3   Hurricane Visualization

We applied our novel illustration-inspired techniques to the application domain of hurricanes. Hurricanes presented us with large amounts of multi-attribute time-varying data as well as with visualization challenges regarding effective representations to convey the time-varying nature of the hurricane for investigative purposes. The direction of motion, the rate of change of intensity of a hurricane, the presence of certain phenomena such as *mesovortices* or *vertical wind shear*, and their effects on a hurricane were important to domain experts. We investigated the use of illustration-inspired techniques in the field of hurricane visualization. In this section, we discuss some of the previous work done in the field of visualizing hurricanes.

Research in hurricane simulation and visualization has typically been conducted using super-computers running simulations of hurricanes for days. One of the first such simulations involved the visualization of hurricane Diana that struck the coast of North Carolina in 1984. The visualization was conducted on a massively parallel IBM supercomputer where the model used 552 processors for two days and produced 100GB of output data (Davis and Bosart 2001), (Davis and Bosart 2002). The research focused primarily on studying the hurricane simulations and obtaining more information regarding the processes behind hurricane evolution.

Weather and climate researchers use a combination of tools that produce one- and two-dimensional visualizations of their data. The Grids Analysis and Display System (GrADS) (COLA 1988), developed at the Center for Ocean, Land and Atmosphere Studies (COLA), is one such 2D visualization tool widely used by atmospheric physicists and weather researchers to explore and study their data. A commercial package widely used by the meteorological community is IDL (Bowman 2005). It is mainly used for its data processing and statistical analysis capabilities, but the newer versions feature some basic three-dimensional graphics functionality. Vis5D (Hibbard and Santek 1990) and VisAD (Hibbard 1998) (the Java version of Vis5D) are some of the most popular applications among weather researchers.

Hurricane visualization has been a topic of particular interest for the last few years because of the immense destruction that has been caused by hurricanes. Hurricane Isabel data

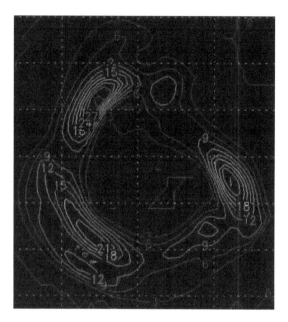

FIG. 2.6. This image shows a standard two-dimensional visualization of isocontour data for cloud water data in hurricane Katrina. The visualization shows information only at a particular altitude from the sea surface and is limited at providing structural information for the entire hurricane.

was made available as part of the IEEE Visualization 2004 contest (Kuo *et al.* 2004). Researchers adopted approaches based on immersive visualization (Gruchalla and Marbach 2004), multithreaded OpenGL-based glyph visualization (Johnson and Burns 2004), interactive brushing in data views (Doleisch *et al.* 2004), anisotropic diffusion for visualizing vector fields (Helgeland and Elboth 2005), variants of volume rendering techniques (Jiang *et al.* 2004), and texture advection techniques for flow visualization (Schafhitzel *et al.* 2004). These techniques, though effective in visualizing certain aspects of the hurricane, were limited in their ability to show the internal structural details of a particular timestep or to convey the temporal nature of the hurricane.

Woodring and Shen (2006) and Sauber *et al.* (2006) discussed new techniques to visualize multi-attribute data, using the multi-attribute hurricane datasets as an example. Their work discussed novel ways to provide insight to domain experts by highlighting attribute correlations between various attributes (such as temperature, pressure, and vorticity) in a single timestep. Such systems, which can help experts find correlations between multiple attributes, can be of immense use to domain experts.

Similar to traditional visualization techniques for time-varying data, the hurricane visualization tools and techniques generate visualizations that merely provide a visual representation of the underlying data without considering the need to highlight, accentuate and emphasize important features in these large multi-attribute datasets. Providing temporal context to increase the understanding of the origin of the storm and the direction of motion in which the hurricane is moving is crucial. Within a single timestep, identifying and accentuating structural features is also of immense use to hurricane experts. An in-depth study evaluating the effectiveness of such techniques from the point of view of domain experts would be very valuable.

Analysis of the data being visualized reveals crucial structural and temporal characteristics that can be used to produce insightful visualizations as well as to annotate visualizations to be more informative. Illustrators distill the characteristic features of their subject and produce illustrations that capture those details, abstracting out irrelevant information. Using techniques that have been inspired by such insightful illustrations, we produce expressive visualizations of hurricanes.

2.4   Visualization of value variant data

Value variant data is time-varying data in which a certain attribute is changing over time (such as global temperature measured over a time interval or infant mortality in the nation over the last eight years). In most cases, the data is spatially two-dimensional and varies over time. Such data is called *2D+t* data and can be visualized using a variety of techniques.

Visualizing such time-varying data is traditionally done by generating a snapshot for each timestep based on a color scale and viewing all the snapshots simultaneously. In such cases, it is extremely easy for a user or even a well trained expert to miss increasing/decreasing patterns as well as interesting variations in values of the attribute being visualized. In some cases, animation of the snapshots can be used as an alternative, but there are always cases of inattentional blindness, where our visual system misses important features due to other irrelevant distractions that cause the viewer to lose focus or attention.

FIG. 2.7. This set of images depicts infant mortality in the entire country over a period of five years from 1998-2002. The color scale provided can be used to understand the data. Identification of clear, interesting patterns and unusual behavior in the data cannot be clearly seen by looking at these snapshots. *Data credits: US Centers for Disease Control and Prevention, National Center for Health Statistics.*

Figure 2.7 shows a visual representation of infant mortality in the nation over a five-year period from 1998 to 2002. The snapshots of the data are limited in their ability to convey interesting patterns in the data. Even simple questions such as which state consistently has the highest infant mortality in the country over the five-year period require the viewer to closely examine each of the images. Using the color scale and their visual ability, a viewer can eventually find the answer. The question, however is not whether such a straightforward question can be answered, but whether more interesting patterns in the data can automatically emerge by using such simple techniques.

2.5   Art-inspired visualization of value variant data

Since traditional snapshot and animation-based techniques are limited in conveying patterns and visualizing change in attributes, we investigated novel techniques from the field of art, in particular from the field of pointillist painting, as invented by Seurat.

Non-traditional computer graphics research has led to interesting imagery in the form of sketch-style images (Winkenbach and Salesin 1994) and painting-style images (Hertzmann 1998), given a photograph or scene as an input. Researchers have developed systems that generate painterly rendering (Haeberli 1990), (Litwinowicz 1997) and animations in a painterly style (Meier 1996). Hertzmann developed a system to generate painterly rendered images given an input photograph in various styles such as impressionism, expressionism, pointillism, and colorist wash (Hertzmann 1998).

Researchers in data visualization have identified the strengths of the paint-based medium and applied them to visualize various kinds of data. Laidlaw *et al.* (1998) used concepts from painting to visualize tensor fields in diffusion tensor imaging. They used oil painting concepts such as layering and encoding information in the brush direction, stroke frequency, and saturation. Figure 2.8 shows an example of a mouse spinal cord visualization using their painterly rendering techniques. They later extended their method to visualize incompressible flows with multiple values (Kirby *et al.* 1999). In this work, they leveraged the concept of multiple layers in a painting to convey the values of multiple attributes in a visualization.

Healey (2001) investigated the use of the painterly medium for effective visualizations of multidimensional datasets. The idea was to map attributes to painterly parameters such as the path of a brush stroke, its length, and the density of the placed brush strokes in a region. In addition, the amount of paint at the beginning of each stroke and a coarseness factor were also mapped to

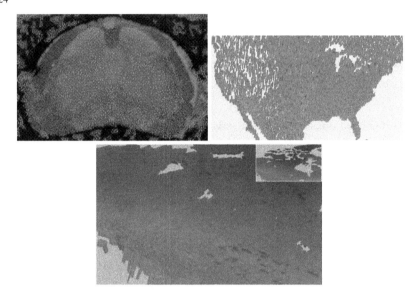

FIG. 2.8. The top left image depicts a painterly visualization of a mouse spinal cord where meaning was associated with attributes of painting such as layering, brush stroke direction, and color saturation (Laidlaw *et al.* 1998). The right image depicts a multi-attribute painterly visualization that shows average weather conditions in the month of August. In this case too, multiple brush attributes such as brush size, color, direction, and coarseness were used to encode information regarding the multiple attributes such as rainfall, temperature, and wind speed into the visualization (Healey 2001). The bottom image is a weather visualization for the month of January for southwest Canada. Aesthetic concepts were included to generate visually pleasing results that would also convey useful information to a viewer (Tateosian *et al.* 2007).

attributes. Perceptual characteristics such as perceptual balance, distinguishability and flexibility of colors were used in addition to the brush stroke attributes. Healey et al. further extended their work in perceptual-based painterly rendering to allow their system to incorporate perceptual knowledge for effective visualization (Healey *et al.* 2004). The top right image in Figure 2.8 shows a visualization of climate conditions using the painterly paradigm. The visualization depicts the mean conditions for the month of August for the United States. Healey *et al.* conducted a rich set of user studies to evaluate their system on paintings as well as multidimensional weather data. The user study results indicated that users preferred the non-photorealistic visualizations

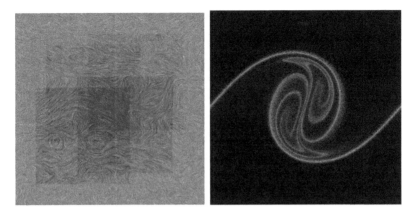

FIG. 2.9. The left image shows a visualization of multi-valued vector data that uses line integral convolution (LIC). The multi-valued vector data is visualized with separate colors, allowing for visual blending, instead of coloring it with a blended color (Urness *et al.* 2003). The right image shows a nanoparticle visualization where the structure is shown through the use of pointillism (Saunders *et al.* 2005).

better than the standard visualizations. Such user studies indicate the strength of such media and emphasize the need for visualization researchers to continue investigating art-based techniques for visualization purposes. Recently, Tateosian *et al.* (2007) have further extended the painterly concept by incorporating artistic concepts such as interpretational complexity (IC), indication and detail (ID), and visual complexity (VC) into the process of generating a visualization. The bottom image in Figure 2.8 shows a visualization of conditions in January over the southwest Canadian region where temperature, pressure, wind speeds, precipitation, and other attributes of weather are mapped to brush, color, and other painting-related attributes.

Urness *et al.* (2003) discussed the concept of sampling and color mixing for visualizing 2D vector distributions. To visualize multi-valued flow data, they applied the pointillism-based principle of visual mixing of colors as opposed to using a resultant color in the visualization. The left image in Figure 2.9 shows an example of a visualization using LIC being depicted using distinct colors instead of blending the colors. Later, Saunders *et al.* (2005) applied concepts from pointillism to visualize nanoparticles in formation. They used pointillism to better visualize the distribution of nanoparticles in space. They augmented their techniques by using glyphs with more information regarding the nanoparticles. The right image in Figure 2.9 shows an example of nanoparticles being visualized using the pointillism paradigm. The figure clearly shows the structural distribution of the nanoparticles and allows users to understand the nanoparticle formation process better.

Chapter 3

ILLUSTRATION-INSPIRED TECHNIQUES FOR
VISUALIZING TIME-VARYING DATA

Visualization of time-varying data has been a challenging problem due to the nature and the size of the datasets. The naïve approach to visualizing time-varying data is to render the three-dimensional volume at each individual time step using standard volume rendering techniques (Cabral *et al.* 1994). This technique relies heavily on the user's ability to identify and track regions of interest over time. At the same time, the number of snapshots generated can be quite high (100-2000) (Davis 2004), requiring considerable effort for the user to track features. To reduce this effort, we draw inspiration from the illustration literature to enable us to convey change over time more succinctly. *Time-varying data* are three-dimensional snapshots of a process captured at regular time intervals. Domains such as computational fluid dynamics, weather forecasting, and medical scans (ultrasound) generate time-varying data.

Time-varying data visualization generally consists of three steps. First, the dataset must be analyzed to identify interesting features. *Features* are regions of interest depending on the scientific domain. Feature extraction can be done manually (where a user selects features), semi-automatically (where an algorithm identifies features that are validated by a user), or automatically (where an algorithm identifies features by analyzing different time steps without user intervention).

The second step in visualizing these datasets is *feature tracking*. The extracted features (from different time steps) are tracked over the time steps. Feature tracking requires the ability to identify the features and correlate them over time.

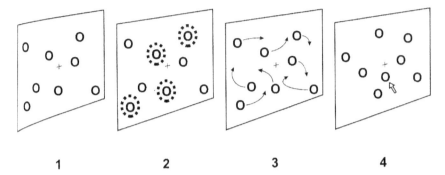

FIG. 3.1. A schematic depicting the multiple object tracking experiment conducted by Pylyshyn. The subject is first shown eight identical objects (panel 1) and then a subset of four are briefly highlighted (panel 2). Next, the objects start moving in a random manner (panel 3), once they stop, the subject is asked to identify the objects that were designated earlier as targets. They found that users were successfully able to track no more than five objects (features) over time (Pylyshyn 2003).

The third step is visualizing this tracked information along with the actual time-varying data. The visualization conveys the change over time in the underlying data. The feature extraction and feature tracking approach is fairly common in time-varying dataset visualization (Silver and Wang 1996).

Generally, identifying and visualizing features over time is a particularly hard task, even for the well trained eye. The problems are numerous, ranging from the large number of time steps that are rendered to the ability of the user/viewer to identify and visually track a particular feature of interest. There is also the problem of occlusion of the feature of interest by another uninteresting feature. The task is further complicated by issues such as the large number of features in each time step.

Pylyshyn (2003), found that observers can track a maximum of five independently moving objects at the same time, as shown in the schematic in Figure 3.1. As the speed of the moving objects and the number of objects increases, the performance of the observers drops considerably. The results of this experiment are particularly significant because they imply that even a trained domain expert will likely not be able to visually track more than five features over time. It is not uncommon to have 10-20 features in a particular dataset; the experimental results imply that it would be impossible for the human visual system to track their paths over time.

FIG. 3.2. This illustration depicts the motion of a bird in flight with the abstract path traversed by the bird and intermediate position indications (MacCurdy 1954).

Illustrations are able to convey information easily, drawing the viewer's attention to the important details and abstracting out the irrelevant details. Illustrators have used numerous techniques to depict change over time in a single image (McCloud 1994).

Leonardo Da Vinci, in his treatise on the the flight of birds (MacCurdy 1954), writes, "The lines of the movements made by birds as they rise are of two kinds, one which is always spiral in the manner of a screw, and the other is rectilinear and curved. That bird will rise up to a height which by means of a circular movement in the shape of a screw makes its reflex movement against the coming of the wind and against the flight of this wind, turning always upon its right or left side." To illustrate this effect, he drew the illustration reproduced in Figure 3.2, which depicts the motion of the bird in flight. The illustration conveys the motion to the viewer by using lines to approximate the path taken by the bird to ascend into the sky. The lines do not connect exact positions of the bird as it took flight, but show an abstract representation that captures the motion in a simple, easy-to-understand manner. Illustrators tend to rely more on abstraction than accuracy when conveying change over time.

The problem of efficiently and succinctly displaying a minimal set of images that conveys information about the interactions within a time-varying dataset is still unsolved. I have identified four techniques from the illustration literature and applied them to depict change over time in time-varying datasets. *Speedlines* are one such technique: that follow features, drawing the user's attention to regions of interest for improved visual tracking of those features. A group of speed-lines together form *flow ribbons*, which convey change of a feature over time more succinctly. Researchers in the field of visualization have used opacity-based techniques to draw the user's attention to a particular feature in a visualization (Silver and Wang 1996). We extend these tech-niques to vary the opacity of features as they transform over time. Opacity-based techniques can be used in conjunction with speedlines to convey positional change better than just using snapshots or animations. Strobe silhouettes is the fourth technique that conveys directional information to the viewer in the form of trailing silhouettes.

3.1 Approach

We first preprocess the data at each time step to identify features of interest. In the second step, the identified features in each timestep are correlated with each other between subsequent time steps to facilitate feature tracking. In order to identify tracks for a feature, we calculate the centroid of the feature and track it over time. Since the centroid of a feature can fall outside the feature, we also track the extreme points of the volume at each time step to provide us with more information for tracking. The centroid and the extreme points are used by our techniques to convey the direction of motion of the feature over time.

3.1.1 Feature identification

We have used feature-extracted data from the Rutgers data repository (Fernandez and Silver 1998). Every voxel in each time step has an identifier that indicates the flow feature at every time step, making it possible to track a particular set of *flow features*.

In this process of enhancing the rendering to draw the user's attention to regions of interest, we first need to identify features of interest, which we shall call *illustration features* to avoid confusion with actual three-dimensional flow features in the data. Illustration features can be features of interest from an application domain experts point of view. Stable, persistent features might be more interesting to some expert whereas some other expert might be interested in visualizing unstable, short-lived features. This can be captured using "illustration features."

In the preprocessing step, we analyze the time-varying data set to identify flow features that are most actively moving (unstable), mostly stable as well as significantly larger compared to its surrounding features. This preprocessing allows us to target a certain class of flow features (stable/unstable) that facilitate more effective visualizations. This selection of features is domain specific.

We define a quantity called the *temporal variation* which measures the amount of change that a feature undergoes over time. The temporal variation is defined in terms of the *coefficient of variation (COV)*, a statistical measure of the standard deviation of a variable. The coefficient of variation has been used for transfer function generation in time-varying data (Jankun-Kelly and Ma 2001), for accelerating volume animation (Shen and Johnson 1994), and for accelerating the rendering of time-varying data using TSP trees (Shen *et al.* 1999), (Ellsworth *et al.* 2000).

The COV for a data value is given by:

$$c_v = \frac{\sigma_v}{\bar{o}_v}$$

where

$$\sigma_v = \sqrt{\frac{1}{n-1}\sum_t (o_{v,t} - \bar{o}_v)^2}$$

and

$$\bar{o}_v = \frac{1}{n}\sum_t o_{v,t}$$

In the above equation, $\bar{o}_v$ is the mean of the sample $o_{v,t}$ under consideration over n time steps, and σ_v is the standard deviation of $o_{v,t}$ from its calculated mean. The COV c_v is calculated by dividing the deviation σ_v by the overall mean $\bar{o}_v$. A larger COV corresponds to high variation and less stability; similarly a smaller value corresponds to more stability over time. In our case, we chose to focus on more stable features, identified by low COV values, for visualization purposes.

Temporal variation is computed by comparing consecutive time steps. For each pair of con-

32

FIG. 3.3. This image depicts the motion of a feature in an experimental data set over time. The direction of motion of this feature is not clear from these snapshots.

FIG. 3.4. This figure shows a set of snapshots from successive time steps of a volume. As is evident from looking at the snapshots, it is very hard to correlate and track a particular feature over different time steps. The problem lies in the fact that many features are very short lived which is why we used the coefficient of variation measure to identify stable, interesting features (illustration features).

secutive time steps, we compute a gradient volume that contains temporal gradients for each voxel. The temporal COV is then computed for the temporal gradients. Temporal gradients give a sense of how the voxel density changed over time for a particular voxel.

We use a synthetic dataset to show the efficacy of the techniques and then present our results on actual CFD data. In Figure 3.3, the feature is moving in a circular manner, but it is not at all apparent from visualizing the individual time steps as shown in the figure. Figure 3.4 shows the snapshots for five consecutive timesteps in the turbulent vortex dataset (Fernandez and Silver 1998). It is evident from these figures that visually tracking a particular feature over time is hard.

3.1.2 Speedlines

Speedlines are defined as lines that convey information to the user about the path traversed by a particular feature over time. They are basically lines that follow a particular feature over time. Illustrators have used speedlines to convey motion by altering the characteristics of these lines. The thickness, line style, and variation of the line's opacity are among the characteristics

FIG. 3.5. This illustration shows the use of speedlines to depict motion of the pitcher in a single frame. Illustration provided courtesy of Kunio Kondo (Kawagishi *et al.* 2003).

FIG. 3.6. This illustration uses speedlines to depict the running motion of the man. The lines get thinner and lighter as they approach the running man. Illustration provided courtesy of Harper-Collins publishers and Scott McCloud (McCloud 1994).

that successfully convey change in direction.

For example, in Figure 3.5, the illustrator successfully conveyed the motion of the pitcher's arm to the viewer (Kawagishi *et al.* 2003). In particular, the thickness of the lines is varied to show the direction of motion of the pitcher's hand. Note that the curve tracing the pitcher's arm is smooth, not irregular: it is an abstraction of the actual movement of a pitcher's arm. The motion of the baseball towards the viewer is successfully depicted by the speedlines that start out thin to depict the origin of motion, thicken to imply the increased intensity during the release of the ball, and then thin towards the end as the ball moves closer to the viewer.

Illustrators use thicker, denser lines to represent older time steps and lighter, thinner lines to represent newer time steps. Figure 3.6 shows an illustration with speedlines. The lines are thicker and darker farther away from the man; they become thinner and lighter as they approach the man.

In Figure 3.7, I have identified one feature and conveyed its motion over twelve timesteps. To convey the notion of time, we have used speedlines. The darker, thicker regions of the line convey an older time step whereas the lighter, thinner regions of the speedline depict a more recent time step. Figure 3.7 enables the viewer to understand the rightward motion of a feature. The

FIG. 3.7. The image shows the use of speedlines to depict motion of a feature in a synthetic dataset through twelve timesteps. The translatory motion of the feature from left to right is depicted using speedlines. The speedlines get thinner and more translucent as they get closer to the latest timestep, just as in Figure 3.6.

FIG. 3.8. The image depicts change over time in a feature for CFD data. The downward, rightward motion of the flow feature is conveyed using speedlines.

characteristics of the speedlines are similar to that of Figure 3.6. The line style is thicker and more opaque in older time instants and thinner and more translucent towards the newer time instants.

We applied the technique of speedlines to real-world data from the computational fluid dynamics (CFD) domain to depict the direction of motion of a particular feature, as shown in Figure 3.8. The downward, rightward motion of the feature is clearly conveyed to a viewer looking at this image.

Researchers in the field of CFD simulation often use particle traces to visualize the exact path taken by a feature. *Particle traces* connect trajectories of a particle over time (Gallagher 1994).

FIG. 3.9. This set of images highlights the difference between particle traces and the speedlines technique. The leftmost image depicts a particle trace of the feature; the middle and rightmost image depict its motion using speedlines. In the middle image, I used every other point from the path traced by the feature and in the rightmost image, I used one out of every four points to generate speedlines.

Particle traces are, by definition, required to be faithful to the path followed by the feature. A speedline, on the other hand, is an expression of how an illustrator would depict the same change. The speedlines approximate the path traversed by a feature and incorporate the smooth, natural strokes of an illustrator to depict the motion of the feature. Speedlines also differ from particle traces in that the line properties such as thickness and opacity of the speedlines are varied to depict the temporal change.

A *streamline*, by definition (Gallagher 1994), is a line that is tangential to the instantaneous velocity direction; generally a collection of streamlines are used by CFD researchers to convey flow. Speedlines, on the other hand, are used to track the motion of a feature over a certain interval of timesteps. The goal of using speedlines is not to convey flow for the entire time-varying dataset, but to facilitate tracking a feature of interest.

Figure 3.9 illustrates the difference between using particle traces and speedlines. The leftmost image depicts a particle trace of the feature: the middle and rightmost image depict its motion using speedlines. In the middle image, we used alternate points from the path traced by the feature; in the rightmost image we used one out of every four points to generate speedlines. Our speedline images are similar to the ones that illustrators would draw to convey the motion of the feature. The speedlines are smooth and succinctly convey the direction of motion to the viewer.

The centroid of the feature in each timestep provides points that are connected using Catmull-

36

Rom interpolating splines. Knowing this information, an offset is calculated in both directions from the centroid curve to simulate the speedlines in an illustrative style. In order to convey the direction of motion, only two speedlines can be used in which case the offset from the centroid line should not extend beyond the extremities of the feature. This ensures that the speedlines are perceived correctly by our visual system as being visual cues pertaining to the feature of interest.

When augmenting a visualization of a larger feature, using more than a single pair of speedlines is more effective, similar to those in Figure 3.6. In such cases, equidistant offsets from the centroid curve are picked for pairs of speedlines.

3.1.3   Flow ribbons

Flow ribbons are used extensively by illustrators to show motion. Flow ribbons are particularly interesting because they occlude underlying regions to depict change. This facilitates the depiction of motion over time.

FIG. 3.10. The illustration depicts the motion of the monster's hand using a flow ribbon. The region near his thighs is occluded and is abstracted using small lines within the flow ribbon. The small lines convey to the viewer the presence of structure under the flow ribbon. Illustration provided courtesy of HarperCollins publishers and Scott McCloud (McCloud 1994).

For example, in Figure 3.10, the illustrator has occluded parts of the monster's legs to depict the motion of his hand. At the same time, the small line segments, inside the flow ribbon (near his

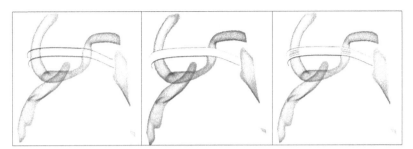

FIG. 3.11. This set of images highlights the difference between various types of flow ribbons. In their simplest form, flow ribbons are merely pairs of speedlines that convey the change to the viewer. In the middle figure, the other extreme in which the region under the flow ribbon is occluded by the ribbon to draw the viewer's attention to the moving feature and emphasize its motion. In the rightmost figure, techniques similar to Figure 3.10 were used to generate the ribbons. Small line segments were used to abstract features occluded by the flow ribbons.

legs), serve as an abstraction to represent a simplified structure of the region of the legs occluded by the flow ribbon. The line segments are drawn by obtaining offsets from the centroid curve. The offsets were picked empirically, but choosing equidistant offsets works equally well.

To obtain flow ribbons, I identify the centroid and the extremities of a selected feature in every time step, and then use that information to draw the flow ribbons. An important characteristic for flow ribbons is that they fade into the background and stop short of the feature. As can be seen in the images in Figure 3.11, this effect gives the viewer an opportunity to mentally complete the diagram by filling in the details. In this process of mentally completing the picture, the viewer is convinced of the change over time.

We define three different types of flow ribbons based on their complexity. Figure 3.11 shows examples for these three types of flow ribbons. The leftmost image shows the simplest type of flow ribbons, in which a pair of speedlines are considered together to convey change over time. The second type of flow ribbons, as shown in the middle image in Figure 3.11, are opaque occluding underlying features. They serve to draw the user's attention to the feature of interest. The third type of flow ribbons use small line segments to abstract the underlying occluded features, as can be seen in the rightmost image in Figure 3.11.

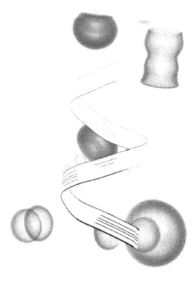

FIG. 3.12. The image shows the helical motion of an experimental data feature using flow ribbons. Just as in Figure 3.10, the occluded features are abstracted using small line segments. The flow ribbon gets thinner as they get closer to the newest timestep.

FIG. 3.13. The image depicts random motion of a flow feature using flow ribbons. The feature moves from bottom right to upper left corner. The features occluded by the flow ribbon are abstracted using thin, small lines to represented the underlying features.

To obtain the line segments overlapping underlying features, an alpha test followed by a stencil test in OpenGL is used. The alpha test checks for underlying features and the stencil test draws the line segments on the underlying feature to provide an abstraction for that feature. The line segments are dynamically generated as the timestep is being investigated. In Figure 3.12, the helical motion of an experimental data feature is shown using flow ribbons. The flow ribbons depict a change of motion along the path of the ribbons. As in Figure 3.10, in regions where the flow ribbon occludes actual data, the line segments convey an abstract representation of their underlying presence.

We applied the flow ribbon techniques to CFD data to track a feature in Figure 3.13. The ribbons occlude the underlying feature and provide an abstracted representation of it by small line segments, similar to Figure 3.10. The motion from the bottom right part of the figure to the upper left part is shown using flow ribbons.

3.1.4   Opacity modulation

Illustrators often used blurred, desaturated images to depict older time steps with brighter, more detailed images representing newer time steps. In visualization, the same effect can be obtained by using opacity modulation techniques.

FIG. 3.14. This illustration conveys the direction of motion of the hand over time. They use low detail, thin lines for the older instants of time and high detail, darker lines for the latest positions of the hand to convey the motion. Illustration provided courtesy of HarperCollins publishers and Scott McCloud (McCloud 1994).

For this technique, an illustration feature is identified and then snapshots of each timestep are merged into one image. At the same time, I modulate the opacity of the older timesteps, making them less opaque and more transparent; the newer timesteps are more opaque and the colors of the

FIG. 3.15. The image conveys change over time using a combination of opacity-based modulating and speedlines. The translatory motion of the feature from upper left to bottom right is conveyed by the image.

newer steps are brighter compared to the older time steps. This provides insight into the origin of the feature and its path through multiple timesteps. We found that the opacity-based techniques in conjunction with speedlines were a better combination to convey the change over time than opacity-based techniques.

Figure 3.14 conveys change over multiple timesteps to the viewer. The varying line thickness and increasing level of detail conveys the left-to-right motion of the hand.

We combined this technique with our speedlines technique, yielding Figure 3.15. The older time step is less saturated and dull, whereas the newer time step is brighter and more well defined compared to the blurred older time steps. This figure conveys the motion of the feature, from the left upper corner to the right bottom corner, to the viewer using a combination of the two techniques very effectively. Just as with the speedlines technique, the thickness of the older line decreases as the line gets closer to the newer timestep.

3.1.5   Strobe silhouettes

Illustrators have used strobe silhouettes to convey previous positions of an object. As can be seen in Figure 3.16, the direction of motion of the axe is apparent from the trailing silhouettes. The strobe silhouettes are increasing in their level of detail as they get closer to the current position of the object. The oldest time step has the most abstract, lowest level-of-detail silhouette. The trailing silhouette effect convincingly conveys the motion of the axe.

FIG. 3.16. The image shows strobe silhouettes depicting motion over time. The downward motion of the axe is conveyed using strobe silhouettes. Illustration provided courtesy of Kunio Kondo (Kawagishi *et al.* 2003).

To obtain strobe silhouettes, we precompute a direction-of-motion vector for the feature. The dot product of the direction-of-motion vector with the gradient vector of the voxel under consideration, as shown in the equation below, identifies whether that voxel should be included in the silhouette computation or not. We combine the silhouettes to get the strobe silhouette effect.

$$(\nabla f_n \cdot motionvector) < 0 \implies Strobesilhouette$$

Figure 3.17 shows an application of strobe silhouettes to a synthetic dataset conveying rightward motion of the feature to the viewer. The feature starts from the left extreme and translates to its current position.

FIG. 3.17. The image shows the strobe silhouettes applied to a synthetic data set, depicting the horizontal motion of the feature from left to right.

42

FIG. 3.18. The image shows the strobe silhouettes applied to flow data. The strobe silhouettes convey the upward motion of the two features. The direction-of-motion vector enables the generation of trailing silhouettes.

Figure 3.18 shows the upward motion of flow features using strobe silhouettes. The direction-of-motion vector facilitates the generation of trailing silhouettes. The bottom figure is moving upwards and the right part of the top feature is opening up. The direction of motion and interaction between features is understood by the viewer due to the use of strobe silhouettes. This image is much more effective at conveying the motion than the snapshots shown in Figure 3.4.

Strobe silhouettes are extremely effective in conveying the direction of motion to the viewer because they provide an abstraction of past time steps. The viewer can mentally recreate the motion with the help of these strobe silhouettes, which helps in conveying positional change over time.

3.2 Multi-feature tracking

Multiple flow features are typically moving in time-varying data. Therefore, I have annotated visualizations of time-varying data containing multiple features with our illustrative cues. Figure 3.19 shows a visualization where the motion of six features is conveyed using speedlines. The image clearly conveys the temporal history of each feature.

Figure 3.20 shows an example in which the motion of three features in real world data is

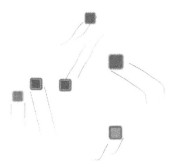

FIG. 3.19. The image shows multiple feature tracking using speedlines for a synthetic dataset with six moving features. In this case, the six features are moving independently of each other and the addition of illustrative cues provides context regarding the direction of motion of each feature.

conveyed using a combination of techniques. Flow ribbons are used to show the motion of a feature and speedlines are used for the motion of the other two features. Since flow ribbons occlude underlying flow features, we use flow ribbons to depict the motion of higher-priority features and speedlines for lower-priority features. The central feature is moving leftwards; its motion is shown using a flow ribbon. The topmost feature's rightward motion and the bottommost feature's leftward motion are shown using speedlines.

3.3   Discussion

The suitability of these techniques is highly dependent on the type of motion that the feature undergoes. Strobe silhouettes an opacity-based techniques are not suitable for types of motion where the feature re-traces the path it has followed, because the silhouettes will overlap each other, making it difficult for the viewer to track the feature and disambiguate older silhouettes with newer ones. Similarly, using opacity-based techniques, an older timestep may be occluded by newer timesteps, which can cause confusion for the viewer. For such motions, speedlines or flow ribbons would be more suitable. Speedlines and flow ribbons techniques would also be suitable for twisting motion. Figure 3.12 depicts spiral motion of a feature using flow ribbons.

This brings us to the use of multiple techniques in a single visualization. We have already

44

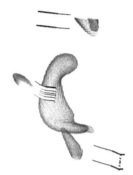

FIG. 3.20. Multiple feature tracking using flow ribbons for the higher-priority flow feature and speedlines for the lower-priority flow features. The central feature's leftward motion is represented using a flow ribbon. The rightward motion of the topmost feature and the leftward motion of the bottommost feature are represented using speedlines.

FIG. 3.21. The rotational motion of the feature in the experimental dataset is conveyed using a combination of opacity-based techniques and flow ribbons. This figure conveys the actual motion of the feature whose snapshots are shown in Figure 3.3. This motion is not at all obvious by looking at the snapshots in Figure 3.3.

seen the combination of speedlines and opacity-based techniques in Figure 3.15. Figure 3.21 uses flow ribbons and opacity-based modulation to convey change over time. This is the actual motion of the feature whose snapshots are shown in Figure 3.3. It is evident that Figure 3.21 conveys the rotational motion of the feature more effectively than the five snapshots in Figure 3.3. The combination of opacity-based techniques and flow ribbons concisely captures the feature's rotational motion.

Chapter 4

INTENT-BASED SIMPLIFICATION OF PATHS

Simplification involves generating a simplified version of a high resolution model. Simplification is increasingly popular in computer graphics, due to the constant need to provide realism as well as interactivity. For example, in a computer game, the participant cares more about the goal of the game than about the details of the surroundings. From a distance, a model containing 100K polygons will look indistinguishable from a simplified representation that contains 1K polygons. Graphics researchers and game developers leverage this to improve the speed and interactivity of the game or application. *Geometric simplification* involves automatically generating a geometric mesh based on the distance of the viewer from the object (Luebke *et al.* 2003). Geometric simplification is also employed to obtain simplified representations of a mesh to reduce the computational complexity or increase the interactivity of an application. The use of this technique is primarily driven by graphics hardware limitations or computational limitations. The challenge in geometric simplification is maintaining visual fidelity and keeping the process of simplification transparent to the viewer. High visual fidelity is hard to maintain, researchers have evaluated simplifications by comparing rendered images of simplified meshes (Lindstrom and Turk 2000) or appearance-preserving simplification based on pixel-based comparison (Cohen *et al.* 1998).

Cognitive simplification, on the other hand, attacks the problem of generating simplified versions using cognition-based principles. The first step in cognitive simplification is to identify features that are important from the point of view of an user. Cognitively significant features are different from perceptually significant features. Perceptually significant features stimulate the visual senses by their ability to capture attention with their color, brightness or co-location with

46

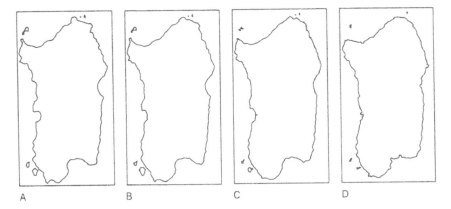

FIG. 4.1. The image above displays increasing simplification of the coastline of Sardinia from left to right. The rightmost image could be used to convey the overall structure of Sardinia whereas for the purpose of coastal shipping, the leftmost detailed map would be more appropriate (Robinson and Sale 1969).

other features. Cognitively significant features, on the other hand, contribute towards the increase of knowledge and assist in the decision making process. They are generally task-specific and the presence of a domain-expert may be required to characterize a cognitively significant feature.

The second step in the cognitive simplification process is the generation of a simplified representation without losing the identified features. The simplification is primarily driven by human factors such as cognitive bandwidth overload and visual clutter. The simplified representation retains the salient characteristics of the underlying features. Ahn and Wohn (2004) have applied techniques to generate simplified representations of motion capture data for scenes with large crowds. They apply simplification techniques to the motion patterns for each character in a large crowd as well as their character mesh for real-time rendering of crowds. In such situations, a cognitive simplification approach would be extremely useful to represent the motion of characters farther away from the viewer in an abstract manner and preserving detail for characters closer to the viewer.

Consider a scenario in which motion capture data of a character consists of a path, as shown in blue in Figure 4.2. Using a naïve one-out-of-n simplification, the loop would be completely eliminated, as shown in the simplified representation depicted in red in the figure. Using more

48

sophisticated geometric simplification algorithms cannot ensure preservation of desired features in a low-level representation. We would like to generate a simplified representation that preserves features in the data. For example, in the right image in Figure 4.2, the simplified representation, shown in red, preserves the loop in the path. A cognitive-simplified representation would be able to preserve such features based on a user-defined function or a mask.

FIG. 4.2. The original path is depicted in blue: A simplified representation obtained by using a simple one-out-of-n simplification algorithm is shown in red in the left image. The right image depicts a feature-preserving representation of the same path using cognitive simplification, also shown in red.

We have applied our path simplification techniques to visualizing the path traversed by the eye of hurricane Katrina. Information regarding the path that the hurricane is likely to take is crucial for planning rescue and evacuation operations. Our cognitive simplification techniques, applied to the path of the hurricane, produce path representations that are easy to use for decision-making purposes, primarily because of their ability to preserve relevant features in the simplified representations.

4.1 Approach

Cognitive simplification is a two-step process. The first step requires the user to specify a function that identifies the features that should be preserved in the simplification process. We call this function the "simplification function." This function could be specified in the form of an equation or could be specified manually in the form of a mask for each point that specifies the simplification factor individually for each point.

The second step consists of performing simplification given the simplification function by automatically identifying the points that should be included into the simplified representation.

4.1.1 Weighted averaging of eliminated points

In the process of simplification, the specified function causes the elimination of certain points of the original curve. The points, though eliminated, are a part of the original curve and contain information related to it that we try to capture using weighting. The primary reason for weighting the eliminated points is to capture the essence of the eliminated points. They could be considered as the equivalent of knots in splines that affect the shape of the curve. Similarly, these eliminated points have user-defined weights that affect the position of the neighboring points included in the simplified curve.

For example, if point i-1 and point i are included in the simplified representation but k points were eliminated between these points due to the simplification function, a weighted average of those eliminated points is used to modify the position of point i. The weighting is Gaussian and is performed using the following formula.

$$s(i) \ = \ \sum_{m=0}^{k} weight(m) * \frac{k-m}{k} * orig\,curve_{i-m} \qquad (4.1)$$

$$weight(m) \ = \ weight - (weight * \frac{k-m}{k}) \qquad (4.2)$$

Equation 4.2 specifies the weight for the eliminated points based on their distance from the current point. The points that are closest to the current point, i, have the largest weight. The weight decreases rapidly as we move away from the current point i. Equation 4.1 specifies the weighting function as the eliminated points from the original curve from i to i-k affect the current simplified point.

In Figure 4.3, the top image depicts, in red, the simplified representation that results from not averaging the eliminated points. The bottom image depicts the effect of weighting eliminated points on the overall appearance of the simplified curve. In the top image, the eliminated points do

50

not have any effect on the shape of the curve especially near the three peaks in the central region of the curve. In the bottom figure though, the simplified representation with weighting clearly affects the shape of the curve by taking into account the eliminated points. The overall shape of the curve, in the bottom image, seems to capture the essence of the original curve better than the simplified representation without the weighting.

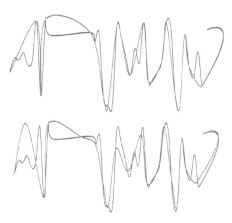

FIG. 4.3. The top image depicts the simplified representation in red but without averaging the eliminated points. The bottom image performs Gaussian weighting on the eliminated points to take those into consideration to affect the overall appearance of the curve. As can be seen in the bottom image, the second bump is a little more tapered and the subsequent bumps too are weighted due to the eliminated points.

Our simplification functions are based on the properties of a curve. We now explore simplification functions that include linearly increasing simplification along the curve, depth-based simplification, simplification based on region-of-interest, and curvature-based functions. We allow the user to specify a simplification function that ensures the preservation of features of interest to the user.

4.1.2 Linearly increasing simplification

Consider a situation where we have information regarding the path traversed by a character in an animation over time. In such a situation, an animator might be more interested in preserving details of the path for the newer instants and simplifying the path for older time instants. In such a situation, a linearly decreasing simplification function could be used to generate feature-preserving simplified representations along the path.

In this method, the simplification factor at the beginning of the curve is maximal. This implies that the simplification is maximum for older time instants and linearly decreases towards the end of the curve where it will be almost faithful to the original curve. The simplification function $s(i)$ is given as

$$s(i) = 1.0 - \frac{i}{n} \tag{4.3}$$

In the above equation, i denotes the index of the current point and n denotes the total number of points in the original curve. As i tends to n, the simplification function tends to 0.

Figure 4.4 depicts the original curve in blue and a simplified representation in red. The simplified representation is obtained by linearly increasing the simplification factor as it approaches the endpoint of the curve. A graph showing the ramp simplification function is shown below in Figure 4.4. As the simplification factor increases, the simplified representation gets less faithful to the original curve. A linearly increasing simplification $s(i)$ could be specified as

$$s(i) = \frac{i}{n} \tag{4.4}$$

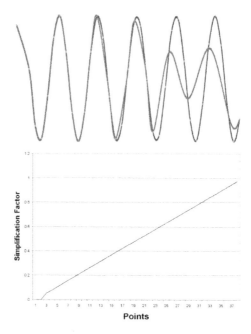

FIG. 4.4. The image depicts the original curve in blue and the simplified representation in red. The ramp simplification function implies increasingly simplification as we traverse the curve. The graph depicts the simplification function as the simplification factor increases towards the end of the curve.

4.1.3 Depth-based simplification

Another scenario from the field of computer animation is that a character running into the distance away from the viewer. In such situations, it would be appropriate to preserve information regarding the path traversed by the character when closer to the viewer, generating a simplified representation relative to its distance from the viewer. The viewer will notice detail in the path closer to them and the depth-based simplified representation will provide the necessary cues to convey the direction of motion of the original path.

In this method, the simplification factor is based on the distance of the points in the curve from the viewer. The simplification factor is high for points that are farther from the viewer, whereas those points close to the viewer have a very small simplification factor. This method preserves detail for features that are closer to the viewer and generates simplified representations for features that are far away from the viewer. The depth-based simplification function $s(i)$ is given by:

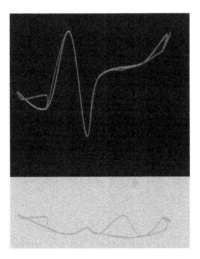

FIG. 4.5. The image depicts the unsimplified curve in blue with its shadow being cast on the surface below. The simplified representation of the curve is depicted in red. The shadow of the simplified curve clearly depicts that points on the curve farther away from the future are simplified and points closer to the viewer are retained almost faithfully.

$$s(i) = \frac{point.z - znear}{zfar - znear} \tag{4.5}$$

where point.z refers to the view space depth of the point under consideration. This simplification function is defined based on the distance of the *z-coordinate* of the current point from the viewer. *Zfar* and *znear* are the distances to the far and near clipping planes respectively; these values are used to normalize the distances to a value in the range [0,1]. As the distance of the current point from the viewer increases, the simplification factor increases and causes more simplification.

In Figure 4.5, the original three-dimensional curve is shown in blue and the depth-based simplified representation is depicted in red. In the orthogonal view, the depth-based simplification is not apparent, but on closely observing the shadow of the simplified representation, the depth-based simplification is obvious. The points closer to the viewer are retained whereas the points constituting features farther away from the viewer are represented in a simplified manner. The preserved detail in points that are closer to the viewer allows the user to explore a particular region without being affected by the surrounding visual clutter.

4.1.4 Distance from center of interest

Consider a situation in which a comet is traveling along a path. Scientists would want to know more about its path information as its proximity to the earth increases. As the comet's path gets closer to the earth, every detail would be crucial, whereas in the other regions, the comet's path would not be as important to the scientists. In such situations, a region of interest (the area surrounding the earth) can be specified; details of the path in that region can be preserved, while the rest of the path may be simplified.

In this case, the simplification factor increases radially as the distance of the points on the curve from the center-of-interest increases. Points that are closest to the center of interest are retained almost faithfully, whereas points constituting features that are farther away are represented in a simplified manner.

$$s(i) = \sqrt{(P_{i_x} - COI_x)^2 + (P_{i_y} - COI_y)^2 + (P_{i_z} - COI_z)^2} \qquad (4.6)$$

In the equation above, P_{i_x} stands for the x-component of the ith point in the original curve and COI_x stands for the x-component of the center-of-interest. The same notation holds for the y- and z-components of the point under consideration and the center of interest. As can be seen from the equation, the simplification factor is larger for points farther away from the region of interest and smaller for points closer to the center of interest.

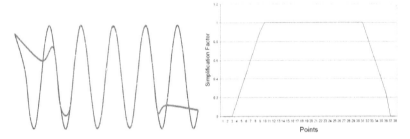

FIG. 4.6. The top image depicts the unsimplified curve in blue and the simplified representation in red. The center of interest is marked in the form of a big green point at the center of the image. A graph depicting the center-of-interest based simplification function explains the situation better.

In Figure 4.6, we see the unsimplified curve in blue and the simplified representation in red. The simplified representation is faithful to the curve in the region surrounding the center of interest, due to the small simplification factor, whereas it is highly simplified in regions farther away from the center of interest.

4.1.5 Curvature-based function

In a situation where one is driving along an unknown road, more information and warning regarding upcoming sharp turns would be helpful to avoid accidents. Naive simplification techniques could completely remove information regarding such a turn, which would be dangerous for

56

the driver. If we have high detail surrounding such regions, it would help the driver to prepare for the upcoming turn. Such sudden, acute turns could be identified automatically using the curvature at each point in the path. We can preserve detail around such regions using curvature-based techniques.

The simplification function is computed based on the curvature of the curve at each point in the original curve. Based on the amount of simplification that is desired, the minimum threshold for the acceptable curvature is specified. The simplified curve contains all the points whose curvature is higher than the threshold.

In Figure 4.7, the blue curve depicts the original curve and the simplified representation of the curve is shown in red. Simplification is performed based on regions that have high curvature. Regions of high curvature could be points of interest and are therefore preserved. The simplified curve is obtained by eliminating points with a low value of curvature since smaller values of curvature generally imply similarity with straight lines.

FIG. 4.7. The original curve is depicted in blue and the simplified representation is in red. High curvature points are preserved in the simplified representation.

4.1.6 Manually specifying a simplification function

In many situations, a domain expert's knowledge and experience are invaluable. Our system has the ability to incorporate such expertise. In the process of simplifying a map, a cartographer might want to preserve path details in tourist spots, since many people are likely to use that detail to navigate. The rest of the path could be simplified more. The cartographer will then specify a mask that preserves path detail at the tourist spots, which will result in a low-level representation with sufficient detail at specified points.

In Figure 4.8, the original curve is depicted in blue and a simplified representation of the original curve is depicted in red. The simplification is performed based on the importance mask specified by the user. A graph of the specified simplification function is shown in the rightmost image. The simplification function can be specified in the form of a file or the user can draw a graph in another application window. As can be seen from the image, the simplified regions of the curve correspond to the peaks of the simplification function graph and the curve is more faithful to the original in the troughs of the simplification function.

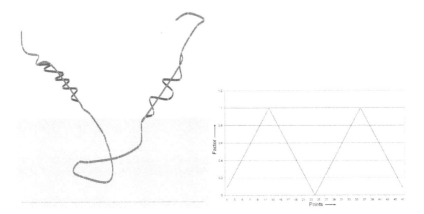

FIG. 4.8. The original curve is depicted in blue and the simplified representation is depicted in red. The simplified representation is based on the manually specified simplification function that is shown below in the form of a graph. The simplified curve is more simplified near the peaks of the simplification function and faithful to the original near the troughs of the function.

4.2 Cognitive simplification of hurricane paths

The cognition-based simplification techniques to the visualization of the path taken by the eye of Hurricane Katrina. *Hurricanes* can be defined as tropical cyclones with high-speed winds. They blow in a large spiral around the central region, which is called the *eye* of the hurricane. A hurricane consists of three main parts: the *eye*, which is a low pressure area at the center of the hurricane; the *eye wall* which is the region around the eye with the fastest winds; and the *rain bands*, which are bands of thunderstorms circulating outward from the eye that feed the storm.

Our dataset is a simulation of the hurricane Katrina and contains attributes such as vertical velocity, relative humidity, temperature, and cloud water. The dataset consists of 70 timesteps, each of which has dimensions $276 \times 288 \times 60$. Figure 4.9 shows a screenshot of the volume rendered image of the 70th timestep. The eye of the hurricane is surrounded by the eyewall. Some rainbands can be noticed towards the right side of the image. We tracked the eye of the hurricane, since it is the region with the highest wind speeds and is representative of the direction of motion of the hurricane. Any other tracking information for the hurricane's direction of motion could easily be used in place of the eye tracking information used here.

FIG. 4.9. The figure depicts a screenshot of the visualization of the hurricane Katrina simulation dataset. The cylinder structure towards the left is the eye of the hurricane surrounded by the eyewall. Some rainbands can be seen towards the right side of the image. The image is a volume rendered image of the cloud water component in the 70th timestep of the hurricane dataset.

Generally hurricanes form over the ocean and sometimes head towards land. When the hurricane is approaching land, information regarding the path that the hurricane could traverse is of critical importance for local authorities. The linearly decreasing simplification function could be

FIG. 4.10. The original path traversed by the eye of Hurricane Katrina.

FIG. 4.11. A ramp simplification function with more detail towards the eye of the current timestep.

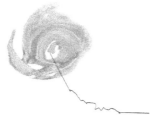

FIG. 4.12. The center-of-interest simplification function that preserves detail around a region of interest.

FIG. 4.13. Curvature based simplification that preserves regions of high curvature.

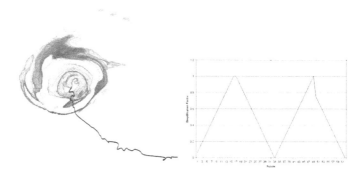

FIG. 4.14. The top image depicts simplification based on a mask specified by the user. The graph shows the manually specified simplification function with two peaks.

used in this case where the simplification factor would be near zero as the hurricane approached land.

Figure 4.10 depicts the original path traversed by the hurricane. A simplified representation using a ramp simplification function is shown in Figure 4.11. The path information contained in earlier timesteps is simplified to convey the general direction.

An important measure of the potential damage and coastal flooding possible due to a hurricane is the *category* of the hurricane. The category of a hurricane is predominantly determined by the speeds of the winds within the hurricane. A category 5 hurricane can have wind speeds of 155 mph. Studying the conditions when a hurricane changes its category is of great interest to meteorologists. In such situations, being able to visualize high detail regarding the path of the hurricane based on a category transition is very useful. A region-of-interest can be specified that will facilitate the study of conditions in these category transition areas. For example, in Figure 4.12, a center-of-interest is identified around the center of the path. The figure shows that the detail in that region is preserved and the path is simplified in the surrounding areas. The region close to the center-of-interest seems to have large perturbations, which that might indicate the transformation undergone by the hurricane. The hurricane seems to be heading downward, but then the intensity of the hurricane increases and it starts moving back up. The zigzag pattern is clearly visible in Figure 4.12, after this section, the hurricane continues moving leftwards and upwards.

High-curvature points in a path can specify unexpected change in direction. Visualizing detail in such regions can help the study of the conditions that caused the abrupt change in direction of the path. In an attempt to identify important features automatically, the curvature at the points constituting the path could be used. High-curvature regions could be used to capture important events in the path of the hurricane. Such regions could be preserved, whereas the simpler, straighter path of the hurricane could be simplified to allow effectively visualizing the path of the hurricane. Figure 4.13 depicts the simplified path that retains detail in high-curvature regions of the path. Salient features near the center of the path and towards the end are well captured in the simplified representation.

A domain expert could specify an importance mask that would indicate regions in which detail needs to be preserved. Use of such an importance mask is extremely useful when regions of interest cannot be defined as clearly by specifying a center-of-interest point discussed earlier. Figure 4.14 depicts such a situation in which a user has specified an importance mask that contains two peaks. The peaks correspond to regions where the simplification factor is highest. In the simplified representation, the mask causes the low-level representation to be faithful to the original path in the central regions as well as the endpoints. In the other regions (corresponding to the peaks), greater simplification is performed.

Our cognitive simplification techniques can greatly help decision-making by presenting more detail regarding the path of the hurricane near critical regions such as heavily inhabited areas, oil rigs and land, in general. The general direction of the hurricane may be sufficient when it is far from the critical areas.

4.3   Semantic simplification of driving directions

Figure 4.15 shows the original path, a naïve simplification and two examples of semantic simplification applied to driving directions. The top image shows the original path, which was obtained from Google Maps for a route from Seattle, WA, to Redmond, WA. The second path from the top shows a simplified representation visualization, which one out of eight points from

62

the original path are selected. It simplifies crucial detail near the source and destination. The third visualization from the top shows a semantic simplification, in which detail is preserved towards the destination. This type of visualization can be useful if one is sufficiently familiar with getting to the nearest interstate from their own house, but needs more detail towards the destination. The fourth visualization from the top shows another example of semantic simplification where the detail along the path is preserved near gas stations (indicated by green circles close to the path). Such a visualization can be useful when one is traveling on a road trip and needs to know where the next gas station is, with some detail around each station.

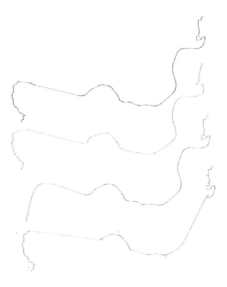

FIG. 4.15. The top image shows the original driving directions path obtained from Google Maps in blue and various simplified paths in red. A naïvely simplified representation is shown in the second visualization from the top. Much detail near the source and destination is lost in this representation. The third visualization shows a semantic simplification in which detail near the destination is preserved whereas the remaining path is simplified. Such a visualization is useful since one is likely to be familiar with the route to the nearest freeway from their own house. The last visualization image shows a path in which detail is preserved near gas stations (shown as green circles near the path).

The hurricane path is obtained by automatic tracking of the center of the eye of the hurricane in every timestep. This provides us with highly detailed information regarding the path of the hurricane. The driving directions is in the GPX data format that is obtained from Google Maps driving directions. Our system allows the user to input any source and destination and converts the data we obtain from Google maps into GPX data points along the path. This data is then used for the path simplification process.

Chapter 5

EVALUATION OF ILLUSTRATION-INSPIRED
TECHNIQUES

This chapter discusses the first of two user evaluations that I conducted to examine the effectiveness of art-inspired techniques. The goal in this particular study was to evaluate the effectiveness of our illustration-inspired techniques in aiding the feature tracking abilities in positionally variant time-varying data visualization. Practitioners currently use large panoramas of snapshots taken over time or watch an animation of these snapshots to perform feature tracking. The aim is to evaluate the benefit of our techniques with respect to user accuracy, time required by the users to complete a task and users' confidence in their answers.

To measure the visual tracking ability of human beings, Pylyshyn (2003) showed subjects a series of moving objects over time. Subjects were asked to visually track these objects as they moved. The study found that a human being can successfully track five objects moving at a relatively moderate speed, but as the speed of the moving objects increased, the user's ability to track the features decreased dramatically. They also found that the user's ability to track features dropped sharply if they were asked to track more than five objects.

5.1   Hypothesis

Our testable hypothesis is that illustration-inspired techniques can lead to improved visual tracking of features as they move over time. By *improved* visual tracking, I mean that our techniques will facilitate faster and more accurate visual tracking than standard snapshot-based or animation-based visualization of time-varying data.

5.2 Independent variables

The independent variables for the user study are the two different ways in which *positionally variant* time-varying data can be viewed:

- A panorama of snapshots versus a single image augmented with our illustration-inspired techniques

- An animation depicting the motion of the feature versus an animation augmented with illustration-inspired techniques.

FIG. 5.1. These are a set of snapshots from successive time steps of a volume. As is evident from looking at the snapshots, it is very hard to correlate and track a particular feature over different time steps.

Figures 5.1 and 5.2 show sample images of a panorama of snapshots and an illustration-inspired visualization of the same data respectively. Inspecting a set of snapshots is a standard visualization technique. However, as can be seen by looking at this series of snapshots, it is very hard to identify the direction in which the contained three-dimensional features are moving. Figure 5.2 shows a sample visualization created using our techniques. The upward movement of the features is clearly conveyed using our strobe silhouette techniques.

To evaluate the illustration-inspired techniques, we compared baseline techniques, such as looking at a panorama of snapshots or a simple animation with visualizations generated by our techniques. There are two primary kinds of illustration-inspired techniques: Speedlines and flow

66

FIG. 5.2. An example of an illustration-inspired technique applied to the vortex data. The strobe silhouettes technique has been applied in this case. Strobe silhouettes convey the upward motion of the two features.

ribbons are similar in nature and opacity modulation and strobe silhouettes are similar in the way they communicate path positions of features. We evaluated our techniques with speedlines and opacity-based techniques, since they were representative of these two classes of illustration-inspired techniques.

5.3   Procedure

Before I began the formal evaluation process, I ran a pilot experiment. The study was not timed or scored for user accuracy.

The pilot study was conducted with three subjects whose answers were not considered in the final evaluation of the techniques. The study revealed some problems, some of which were not obvious, and allowed us to fix them before we began the entire user study (Martin 2003). The problems that were identified were as follows:

- In some cases, there were more than one correct answers depending on the interpretation. This was fixed to ensure that there was atmost one correct answer for the task being performed.

- Subjects were inconsistently interpreting the task they were required to perform. Rewording and expanding task descriptions solved that problem.

- Incorrect options for the task being performed by the subjects. In a couple of cases the subjects could not complete the task, since there was no correct answer for the task they were performing.

- Spelling mistakes in sentences.

5.3.1 Subjects

We tested our techniques with 24 subjects who had basic familiarity with using computers. We did not restrict ourselves to any age group or gender. We performed full factorial, within-subjects testing to evaluate our techniques. In order to balance ordering effects, we tested the subjects with all possible combinations of orderings of trials.

Table 5.1 shows the ordering used for the subjects. Within each category (such as "Snapshots"), we tested the subjects for four different visualization techniques. The ordering used within each category is provided in Table 5.2. Table 5.3 shows the entire table for the study enumerating the order in which each user would see each visualization.

Sub ID	Snapshots	Augmented Snapshots	Animation	Augmented Animation
1	1	2s	3	4o
2	1	2s	4	3o
3	1	3s	2	4o
4	1	3s	4	2o
5	1	4s	2	3o
6	1	4s	3	2o
7	2	1s	3	4o
8	2	1s	4	3o
9	2	3s	1	4o
10	2	3s	4	1o
11	2	4s	1	3o
12	2	4s	3	1o
13	3	1o	2	4s
14	3	1o	4	2s
15	3	2o	1	4s
16	3	2o	4	1s
17	3	4o	2	1s
18	3	4o	1	2s
19	4	1o	3	2s
20	4	1o	2	3s
21	4	2o	1	3s
22	4	2o	3	1s
23	4	3o	2	1s
24	4	3o	1	2s

Table 5.1. Ordering for evaluation of our techniques in comparison to standard visualization techniques. Here **s** stands for speedlines and **o** stands for opacity-based techniques, as discussed in the previous chapter.

a	b	c	d
a	b	c	d
a	b	d	c
a	c	b	d
a	c	d	b
a	d	b	c
a	d	c	b
b	a	c	d
b	a	d	c
b	c	a	d
b	c	d	a
b	d	a	c
b	d	c	a
c	a	b	d
c	a	d	b
c	b	a	d
c	b	d	a
c	d	a	b
c	d	b	a
d	a	b	c
d	a	c	b
d	b	a	c
d	b	c	a
d	c	a	b
d	c	b	a

Table 5.2. In each of the four categories mentioned in table 5.1, the subject is shown four different images/animations. These examples are varied according to the order given above.

Sub ID	Snapshot				Augmented Snapshots				Animation				Augmented Animation			
1	1-a	1-b	1-c	1-d	2s-a	2s-b	2s-c	2s-d	3-a	3-b	3-c	3-d	4o-a	4o-b	4o-c	4o-d
2	1-a	1-b	1-d	1-c	2s-a	2s-b	2s-d	2s-c	4-a	4-b	4-d	4-c	3o-a	3o-b	3o-d	3o-c
3	1-a	1-c	1-b	1-d	3s-a	3s-c	3s-b	3s-d	2-a	2-c	2-b	2-d	4o-a	4o-c	4o-b	4o-d
4	1-a	1-c	1-d	1-b	3s-a	3s-c	3s-d	3s-b	4-a	4-c	4-d	4-b	2o-a	2o-c	2o-d	2o-b
5	1-a	1-d	1-b	1-c	4s-a	4s-d	4s-b	4s-c	2-a	2-d	2-b	2-c	3o-a	3o-d	3o-b	3o-c
6	1-a	1-d	1-c	1-b	4s-a	4s-d	4s-c	4s-b	3-a	3-d	3-c	3-b	2o-a	2o-d	2o-c	2o-b
7	2-b	2-a	2-c	2-d	1s-b	1s-a	1s-c	1s-d	3-b	3-a	3-c	3-d	4o-b	4o-a	4o-c	4o-d
8	2-b	2-a	2-d	2-c	1s-b	1s-a	1s-d	1s-c	4-b	4-a	4-d	4-c	3o-b	3o-a	3o-d	3o-c
9	2-b	2-c	2-a	2-d	3s-b	3s-c	3s-a	3s-d	1-b	1-c	1-a	1-d	4o-b	4o-c	4o-a	4o-d
10	2-b	2-c	2-d	2-a	3s-b	3s-c	3s-d	3s-a	4-b	4-c	4-d	4-a	1o-b	1o-c	1o-d	1o-a
11	2-b	2-d	2-a	2-c	4s-b	4s-d	4s-a	4s-c	1-b	1-d	1-a	1-c	3o-b	3o-d	3o-a	3o-c
12	2-b	2-d	2-c	2-a	4s-b	4s-d	4s-c	4s-a	3-b	3-d	3-c	3-a	1o-b	1o-d	1o-c	1o-a
13	3-c	3-a	3-b	3-d	1s-c	1s-a	1s-b	1s-d	2-c	2-a	2-b	2-d	4o-c	4o-a	4o-b	4o-d
14	3-c	3-a	3-d	3-b	1s-c	1s-a	1s-d	1s-b	4-c	4-a	4-d	4-b	2o-c	2o-a	2o-d	2o-b
15	3-c	3-b	3-a	3-d	2s-c	2s-b	2s-a	2s-d	1-c	1-b	1-a	1-d	4o-c	4o-b	4o-a	4o-d
16	3-c	3-b	3-d	3-a	2s-c	2s-b	2s-d	2s-a	4-c	4-b	4-d	4-a	1o-c	1o-b	1o-d	1o-a
17	3-c	3-d	3-a	3-b	4s-c	4s-d	4s-a	4s-b	1-c	1-d	1-a	1-b	2o-c	2o-d	2o-a	2o-b
18	3-c	3-d	3-b	3-a	4s-c	4s-d	4s-b	4s-a	2-c	2-d	2-b	2-a	1o-c	1o-d	1o-b	1o-a
19	4-d	4-a	4-b	4-c	1s-d	1s-a	1s-b	1s-c	2-d	2-a	2-b	2-c	3o-d	3o-a	3o-b	3o-c
20	4-d	4-a	4-c	4-b	1s-d	1s-a	1s-c	1s-b	3-d	3-a	3-c	3-b	2o-d	2o-a	2o-c	2o-b
21	4-d	4-b	4-a	4-c	2s-d	2s-b	2s-a	2s-c	1-d	1-b	1-a	1-c	3o-d	3o-b	3o-a	3o-c
22	4-d	4-b	4-c	4-a	2s-d	2s-b	2s-c	2s-a	3-d	3-b	3-c	3-a	1o-d	1o-b	1o-c	1o-a
23	4-d	4-c	4-a	4-b	3s-d	3s-c	3s-a	3s-b	1-d	1-c	1-a	1-b	2o-d	2o-c	2o-a	2o-b
24	4-d	4-c	4-b	4-a	3s-d	3s-c	3s-b	3s-a	2-d	2-c	2-b	2-a	1o-d	1o-c	1o-b	1o-a

Table 5.3. Complete table enumerating the ordering of visualizations for user evaluation of illustration-inspired techniques.

5.3.2 Datasets

A mix of synthetic and real-world data was used. For initial training purposes, synthetic data is suitable to familiarize the users with the nature of the dataset and tasks that they will be asked to perform.

The desired characteristics of the datasets to be used for evaluation are:

- The data are three-dimensional in nature and contain three-dimensional features. There is at least one feature in each dataset.

- Features are tracked *a priori* to facilitate visualization. This means that our work does not deal with the actual task of identifying and tracking a feature over time. Our work focuses on aiding *visual tracking* of a feature and does not deal with automatic feature tracking in time-varying data (Samtaney *et al.* 1994), (Silver and Wang 1996). The paths traversed by the contained features is provided along with the three-dimensional data.

In order to evaluate our visualization techniques, we developed several synthetic datasets that have the characteristics mentioned above. Since the datasets were generated by hand, they were very useful for evaluating the strengths and weaknesses of a visualization technique.

The synthetic datasets that we used were:

1. Simple linear/circular motion of a feature

2. Multiple features with simple motion paths: linear, circular, and spiral

3. Multiple features moving at the same time; with complex motions

The real world data that we used is the Turbulent vortex dataset from Rutgers University (Fernandez and Silver 1998). The dataset is a pseudospectral simulation of coherent turbulent vortex structures with a $128 \times 128 \times 128$ resolution (100 time steps). The variable being visualized is vorticity magnitude. This time-varying dataset has numerous features that change over time, sometimes splitting into multiple features; multiple features can also merge into a single feature.

During the user study, we followed this procedure:

1. Explain the user study to the subject and inform them of what it entails.

2. Obtain consent from the subjects for the user study.

3. Present them with training material and let them familiarize themselves with the task for some time.

4. Conduct the user study and collect the results after they were done

5. Present a usability questionnaire and get subjective feedback from the subjects. Questions for this questionnaire are specified in the next section.

5.3.3 Subjective evaluation

Subjects were requested to fill out a questionnaire evaluating the usability of the various techniques. Their answers were obtained on a scale of 1 (easy/agree) to 9 (hard/disagree). The questions asked in the questionnaire were as follows:

1. Were the questions asked for each evaluation straightforward?

2. Overall, did you think the speedlines techniques helped convey direction better than standard visualization techniques?

3. Overall, did you think the opacity-based techniques helped convey direction better than standard visualization techniques?

4. Could you perform simple tasks such as tracking a single feature using standard snapshots or animation-based techniques?

5. Could you perform simple tasks such as tracking a single feature using illustration-based techniques?

6. Could you perform hard tasks such as tracking multiple features using standard snapshots or animation-based techniques?

7. Could you perform hard tasks such as tracking multiple features using illustration-based techniques?

5.4 Tasks

We primarily ask users to track features as they are moving over time. Such tasks are representative of what researchers need to do on a regular basis as they track vortex tubes in feature data, as they track hurricane features over time, as they track the energy of jets as they enter a region, and so on. Our tasks were simple enough to test the effectiveness of our techniques with subjects who are not necessarily application-domain experts; nevertheless, the same time they represent tasks that scientists need to perform on a regular basis.

The task-based question that we asked the user was

Which of these paths seems to best represent the observed direction of motion of the feature?

The subject was presented with four choices of paths and was asked to select one of them according to the perceived motion. This enabled us to test complex motion paths instead of just simple linear motion of features.

5.5 Dependent variables

We measure the accuracy of the users in performing the task in addition to the time required to complete the task. The *user performance time* is the time required by the user to read the question, look at the snapshots or animation and then submit the answer. Time helps us analyze whether subjects perform the task faster using our techniques as compared to standard techniques. Additionally, we measured the *confidence* of the user in their answer. We also obtain feedback in the form of *subjective satisfaction* where we ask subjects to rate their experience on a scale of 1-9 (Likert scale).

74

5.6   Conducting the user study using a web browser

The user study was conducted using a web browser. Since we had to show videos to the subjects, I used Riva Free FLV encoder to encode AVI files into flash files that can be shown to the viewer in the browser setting. Figure 5.3 is a screenshot of the first screen shown to the subjects. It gives an overview of the user study to the subjects. After answering any other questions that the subject might have, I request the consent of the subject to continue the user study.

FIG. 5.3. A screenshot of the welcome screen that was shown to the subjects. The user study was explained to the subjects and any other questions were answered before obtaining their consent.

Figures 5.4 and 5.5 show screenshots of screens that the subjects see. We first ask the user a question and depending on the kind of data (snapshots or animation), the user then observes it and provides an answer. We ask the user to specify the direction in which a particular feature is moving.

The user then picks one of the given choices and indicates a confidence level in their answer. We use a Likert scale to measure the confidence that the subjects have in their answers. The user was asked to select a confidence level from one of: Not at all confident, Slightly confident, Confident, Highly Confident, and Completely confident.

FIG. 5.4. A screenshot of a sample screen that is shown to the subjects. The subjects were shown images and were asked to indicate the direction of motion of a particular feature moving over time. They were also requested to give their confidence level in their answer. This helps us identify how confident the subjects were with their answers using standard as well as our illustration-inspired techniques.

5.7   Results

To measure the effectiveness of our techniques, I analyze the results of the user study using statistical techniques. In order to obtain a quantitative comparison of our techniques, we evaluate the accuracy, the time required for subjects to complete a task, and their confidence in their answers. To compare the four different visualization types, I use the statistical test *Analysis of Variance (ANOVA)*. This test allows us to compare the accuracy, timings, and confidence obtained from the four groups (snapshots, augmented snapshots, animations, and augmented animations). The test begins with a null hypothesis that the use of illustration-inspired techniques provides no speedup in completing tasks, no improvement in accuracy, and that the users feel equally confident

FIG. 5.5. This is a sample screen that is shown to the subjects. Here the subjects were shown an animation and were asked to indicate the direction of motion of a particular feature moving over time. They were also requested to indicate their level of confidence in their answer.

in their answers for all techniques. The statistical measure of significance p evaluates the probability of the result agreeing with the null hypothesis. For values of $p < 0.05$, the null hypothesis is rejected, implying that the use of illustration-inspired techniques makes a difference.

The first metric that was used to evaluate our techniques was user accuracy, defined as the number of correct answers per user per technique. The means and standard deviation for each technique are listed in Table 5.4. Figure 5.6 shows a graph of the same data. A comparison of the snapshots technique with the snapshots technique augmented with our techniques shows that the subjects got more answers correct using the augmented snapshots. Similarly, in the case of animations compared to augmented animations, the subjects were more accurate when using augmented animations, as can be seen in Figure 5.6. The temporal context that the illustration-inspired techniques provide seem to help users complete the task more accurately. Amongst all the four techniques, the augmented animations techniques seems to provide users with the most useful information to correctly complete the task.

Analyzing the accuracy results using the ANOVA test yields the results shown in Table 5.7. The variation between the four different visualization techniques is high. The probability p of this

Type of visualization	Mean	Standard Deviation
Snapshots	76.851	15.14
Augmented Snapshots	91.67	13.87
Animations	91.67	12.01
Augmented Animations	97.22	10.59

Table 5.4. This table shows the mean and standard deviation of the accuracy of the users. Users completed tasks with the most accuracy in the case of augmented animations.

result assuming the null hypothesis is less than 0.0001. This implies that the result is *extremely significant* and that the null hypothesis is rejected. This proves that the use of illustration-inspired techniques increases the user accuracy for visual tracking of features.

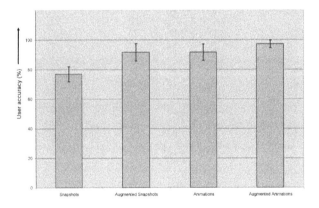

FIG. 5.6. This graph shows accuracy results grouped according to the categories of questions asked. The categories, from left to right, are Snapshots, Augmented Snapshots, Animations, and Augmented Animations. The user accuracy for Augmented Snapshots is better than Snapshots, similarly the accuracy is better for Augmented Animations as compared to Animations.

Additionally, we measured the time required by the subject to complete each task per visualization technique. Table 5.6 shows the mean and standard deviation for the four visualization techniques. Figure 5.7 shows a graphical representation of the timing results. The subjects required more time when viewing snapshots than in the other three cases. Augmented animations helped subjects answer questions faster than just animations. Overall, even though loading an animation

Source of variation	Sum of squares	Degrees of freedom	Mean squares	F
between	6181.0	3	2060.0	12.02
error	1.78E+04	104	171.4	
total	2.40E+04	107		

Table 5.5. This table shows the result of performing the ANOVA test on the accuracy per user. The probability of this results, assuming the null hypothesis, is less than 0.0001. This implies that the result is extremely significant and the null hypothesis that the illustration-inspired techniques do not increase the accuracy in completing a task, is rejected.

Type of visualization	Mean	95% confidence interval	Standard Deviation
Snapshots	54.822	47.85-61.80	2.92
Augmented Snapshots	22.696	15.72-29.67	2.449
Animations	33.411	26.44-40.38	3.107
Augmented Animations	26.248	19.27-33.22	2.539

Table 5.6. This table shows the mean, 95% confidence intervals around the mean and standard deviation for the timing results.

took more time, animations seemed to give the user a better understanding of the time-varying nature of the data.

The time required to complete a task in each of the four cases was analyzed using ANOVA. Table 5.7 shows the result of the ANOVA test on the timing data. The value of p computed using ANOVA was less than 0.0001 which implied that the null hypothesis was rejected. The value of p implies that the result is *extremely significant* according to the ANOVA test and implies that the use of illustration-inspired techniques clearly helped users complete tasks faster than with just snapshots or animations.

In addition to the user accuracy and the time required by the subjects, we requested the sub-

Source of variation	Sum of squares	Degrees of freedom	Mean squares	F
between	1.6779E+04	3	5593.0	16.75
error	3.4724E+04	104	333.9	
total	5.1503E+04	107		

Table 5.7. This table shows the result of performing the ANOVA test on the time required by the users to complete a task using all the techniques. The probability of this results, assuming the null hypothesis, is less than 0.0001. This implies that the result is extremely significant and the null hypothesis that the illustration-inspired techniques do not increase the accuracy in completing a task, is rejected.

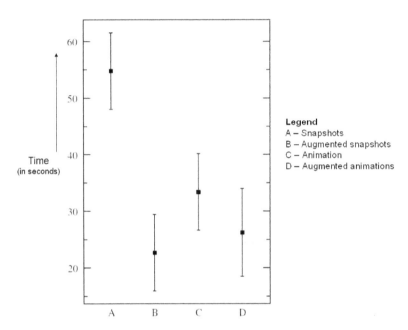

FIG. 5.7. This graph shows the amount of time required by the subjects to complete the task using the four techniques. Users took more time to complete a task using snapshots as compared to all the other techniques. Users also took more time to answer questions using animations as compared to animations augmented with illustration-inspired techniques. Inspite of the fact that the users had to wait for the animation to load, they were able to answer questions faster for animations and augmented animations as compared to plain snapshots.

Type of visualization	Mean	95% confidence interval	Standard Deviation
Snapshots	3.6389	3.418-3.860	0.088
Augmented Snapshots	4.0278	3.807-4.249	0.078
Animations	3.7407	3.520-3.962	0.103
Augmented Animations	4.4907	4.270-4.712	0.094

Table 5.8. This table shows the mean, 95% confidence intervals around the mean as well as standard deviation for the confidence results.

Source of variation	Sum of squares	Degrees of freedom	Mean squares	F
between	40.23	3	13.41	5.806
error	988.5	104	2.310	
total	1029.0	107		

Table 5.9. This table shows the result of performing the ANOVA test on the confidence per user. The probability of this results, assuming the null hypothesis, is less than 0.0007. This implies that the result is statistically significant and the null hypothesis that the illustration-inspired techniques does not increase the confidence of the subjects in their answers, is rejected.

jects to specify a confidence level for each question. Table 5.8 shows the mean and standard deviation of the confidence obtained for each technique. The confidence was higher for both augmented snapshots and augmented animations, as can be seen in Figure 5.8. The subjects had the least confidence in their answers for the plain snapshots and low confidence for the plain animations.

Table 5.7 shows the results of performing the ANOVA test on user confidence per question. Analyzing the confidence that the subjects had in their results, we found that subjects were more confident in the correctness of their answers when using illustration-inspired techniques. The value of p from the ANOVA test obtained was $p < 0.0007$ which according to the ANOVA test is an *extremely significant* result and rejects the null hypothesis that the users feel equally confident with and without illustration-inspired techniques. Our techniques clearly instill more confidence in the users in both the augmented snapshots and augmented animations.

5.8 Discussion

Based on the analysis of the user study, we can say that our illustration-inspired techniques help users track features in three-dimensional data faster and with more accuracy.

For synthetic datasets, the users were faster and much more accurate for all the four visual-

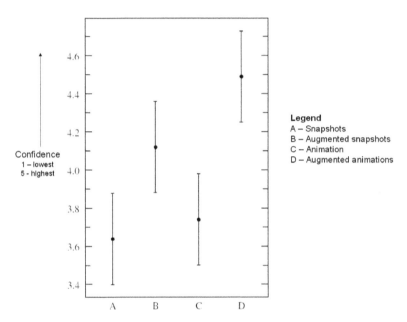

FIG. 5.8. The users were asked to specify their confidence in their answers. This graph shows a representation of the overall confidence that the users had in their answers. As can be seen, users were more confident about their answers for the augmented snapshots and animated animations as compared to plain snapshots or animations.

ization techniques. In the case of real-world data, users were less accurate without the illustration-inspired techniques as can be seen in Figure 5.9. The illustration-inspired techniques when applied to real-world datasets clearly aided the user in accurately identifying the direction of motion of the features.

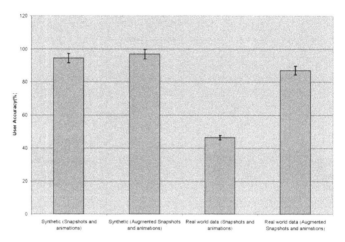

FIG. 5.9. In this graph, the accuracy of the users for synthetic data versus real world data is compared. As can be seen, the accuracy for synthetic data is high for both with and without illustration-inspired techniques. The accuracy for real world data without illustration-inspired techniques is very low whereas the use of illustration-inspired techniques when applied to real world data has boosted the accuracy of the users.

Among the four techniques, the subjects preferred the animations with illustration-inspired techniques the most. This could have been due to the fact that the illustration-inspired techniques augmented animations without distracting the viewer from the main aim of the viewing the time-varying phenomenon. The illustrative cues provide important cues to the viewer in a non-invasive manner and clearly help users to complete tasks faster and more accurately and faster than the other techniques. The ability to complete tasks correctly with the use of illustration-inspired techniques seems to have given the users confidence regarding the correctness of their answers.

Our subjective evaluation gave us feedback regarding the users' preferences for the various techniques. Almost all the users thought that both the speedlines and the opacity-based techniques helped convey direction better than only snapshots or only animations. Of the two techniques, the users indicated through their answers that they preferred the speedlines technique to the opacity-based techniques. From discussions with domain experts, we think this might be due to misinterpretation of older timesteps actually being data for the current timestep. The amount of opacity variation from one timestep to another needs to be investigated further to avoid such confusion.

Users also said that it was easier to perform simple tasks, such as tracking a single feature of interest, using our illustration-inspired techniques. Tracking multiple features of interest using snapshots was found to be very hard, but illustration-inspired techniques were still preferred to simple snapshots or animations for multi-feature tracking.

Chapter 6

CASE STUDY: EFFECTIVE VISUALIZATION OF HURRICANES USING ILLUSTRATION-INSPIRED TECHNIQUES

New visualization techniques are of limited use if not applied to multiple application domains. The use of the developed visualization techniques can be better understood after applying them to several specific application domains. Experts from those domains provide critical insight into the true value of the techniques and can suggest specific scenarios in which those techniques can help them perform a specific task.

In our case, we applied our illustration-inspired techniques to the field of hurricanes. Through a domain expert, we not only had access to hurricane datasets but were also able to generate visualizations of value to researchers in the field of atmospheric physics. Hurricane visualizations generated by using our illustration-inspired techniques, were evaluated by domain experts to provide us with feedback regarding the use of those techniques. They saw value in the techniques and preferred them in most cases over the standard two-dimensional visualizations that they currently use.

6.1 Introduction

Accurate forecasts of the tracks and intensification of hurricanes are crucial in order to minimize loss of life and to plan evacuation strategies. The *category* of a hurricane is an important measure of the potential damage and coastal flooding. The category of a hurricane, as defined by the Saffir-Simpson scale, is predominantly determined by the sustained wind speeds and the minimum central pressure of the hurricane. Figure 6.1 shows a table that specifies how the category of a hurricane is identified based on its sustained wind speeds. The category ranges from the weakest category, 1 (74-95 mph), to the most intense category, 5 (greater than 155 mph).

Category	Central Pressure (mb)	Winds m/s (mph)	Typical Damage
1	>= 980	33-42 (74-95)	No real damage to building structures. Some coastal road flooding and minor pier damage.
2	965-979	43-49 (96-110)	Some roof, door and window damage to buildings. Considerable damage to mobile homes, poorly constructed signs and piers.
3	945-964	49-58(111-130)	Some structural damage to small residences and utility buildings. Low-lying escape routes cut by rising water 3-5 hours before arrival of the hurricane center.
4	920-944	58-69(131-155)	Complete roof structural failures on small residences. Complete destruction to mobile homes, trees and all signs. Major damage to lower floors of structures near the shore.
5	<920	>69 (155)	Roof failure on many residencess and industrial buildings. Massive evacuation of residential areas on low ground within 5-10 miles of the shoreline required.

FIG. 6.1. The Saffir-Simpson scale of hurricane intensity.

Certain structural features of hurricanes, and their evolution in time, give information about the dynamical processes involved in the intensification or dissipation of a hurricane. Common representations are two-dimensional visualizations, which allow the user to visualize the data only at a particular height/level in the data (COLA 1988). This limitation may prevent the user from being able to correlate features on that level with surrounding spatial features or with features in adjacent timesteps. This significantly limits the user's ability to explore the data for further investigation and discovery. Three-dimensional visualization systems can provide an overview of the spatial structure of a phenomenon such as a hurricane (Hibbard and Santek 1990), (Hibbard 1998). Such systems work well for visualizing the overall structure using standard visualization techniques such as isosurfacing, volume rendering and iso-contouring. However, such techniques are limited by the fact that they may occlude internal structural details and are not able to provide temporal context to experts investigating hurricanes.

The process of understanding spatial relationships between hurricane features in a single timestep and temporal relationships within the time-varying data is one of the main challenges in this application domain. Specific challenges deal with the identification and visualization of the vertical wind shear and visualization of the formation and evolution of small scale structures in the hurricane. The hurricane is embedded in the larger scale global wind field. *Vertical wind shear* occurs when the magnitude and direction of the winds surrounding the hurricane vary across the height of the hurricane, causing the storm to lose vertical coherence. A sufficiently strong vertical wind shear will lead to the weakening of a storm. The region surrounding the eye of the hurricane, called the *eyewall*, is closely observed by experts and visualization of attributes in that region immensely helps in the process of understanding and correlating the scientific models to the observed structure in the hurricane.

The use of illustration-inspired techniques, such as volume illustration techniques (Rheingans and Ebert 2001), stippling techniques (Lu *et al.* 2002), importance driven volume visualization (Viola *et al.* 2005), exploratory volume visualization (Bruckner and Gröller 2006), (Correa *et al.* 2006), and flow illustration techniques (Joshi and Rheingans 2005), (Svakhine *et al.* 2005), have resulted in informative and effective visualizations of structural and time-varying data. I use techniques introduced in Chapter 4 and from this field of research to visually investigate and understand the structural and temporal changes in a hurricane.

Drawing on respective expertise in visualization and atmospheric physics, we approached the application domain based on the questions we were attempting to answer. To answer questions related to a single timestep, such as identification and quantification of the vertical wind shear, we used techniques to accentuate hurricane features in a single timestep. In order to answer questions regarding the time evolution process of the hurricane, we used illustration-inspired techniques to display the change in attribute values when the hurricane changes its category. The visualization of temporal information in a single image provides useful semantic information about the rate of intensification of the hurricane.

To evaluate the resulting visualizations, I asked experts and practitioners in the field of hurricane research for critical feedback. I presented them with a set of images that contained two-dimensional visualizations, traditional three-dimensional visualizations and visualizations generated using our illustration-inspired techniques. Experts generally preferred our visualizations over those generated with their standard tools.

6.2 Application domain

A *hurricane* is another name for a tropical cyclone, which is a vertically coherent vortex of rapidly swirling air with a low-pressure core that forms in the tropical Atlantic and Eastern Pacific oceans. Warm moist air over the tropical oceans spirals inward at the sea surface and ascends in the *eyewall*, the cloudy region of maximum winds surrounding the cloud-free *eye*, where the air is descending on average. The air flow is outward at the top of the hurricane, and the cloudy spiral arms are called the *rainbands*.

The inward spiraling air gathers up moisture; as it rises, water vapor condenses into clouds and rain with an accompanying release of latent heat. Latent heat of condensation, the energy source for the storm, leads to further vertical motion, which accelerates the winds via *vortex tube stretching*. Ultimately, the fuel for the hurricane "engine" resides in the warm surface layer of the tropical ocean (NOAA 1999).

Hurricanes form from thunderstorm clusters in environments that exhibit four specific characteristics. The following conditions have to be met for the formation of a hurricane.

- The sea-surface temperature must equal or exceed about 27° C (81° F)

- The surface layer of warm water in the ocean must be sufficiently deep, typically 60 meters or more.

- The winds in the atmosphere must not change substantially with height (i.e., there must be no vertical wind shear).

- The location must be at least five degrees north or south of the equator

While forecasts of hurricane tracks have improved in recent years, accurate prediction of hurricane intensity change has proved elusive. Increases in the spatial and temporal resolution of both satellite observations and NWP models have revealed numerous small-scale, turbulent flow features that may play an important role in determining hurricane intensity, although the exact mechanisms are not known (Montgomery *et al.* 2002). Among these are *mesovortices*, small-scale, intense vortices in the eye and eyewall.

Mesovortices in the eyewall are thought to be associated with strong vertical motion in so-called *hot towers*, and the rising eyewall motion may in fact be largely concentrated in these regions (Braun *et al.* 2006). Other effects of eyewall instabilities include radial transport of warm, buoyant air from the eye to the eyewall, or transport of angular momentum from the eyewall to the eye, which is thought to be important in intensification (Montgomery *et al.* 2002). In addition to their possible role in intensity change, mesovortices may lead to significant increases in damage to coastal areas, due to their high tornado-like winds.

Vertical wind shear has been known to cause dissipation in a hurricane and is closely observed by meteorologists during the evolution of a hurricane. Current methods used by meteorologists to observe wind shear consist of looking at animations of satellite data of the movement of cloud features. A recent analysis of the hurricane Bonnie simulation used to study their effect (Braun *et al.* 2006) suggests that wind shear, convection, and mesovortex formation are linked. One of the objectives in this area of study is to develop methods that allow this connection to be seen clearly in three dimensions.

The goals specific to the application domain included both generating images for educational and outreach purposes and generating insightful visualizations to facilitate exploration and attain a better understanding of the internal workings of a hurricane. Identifying basic structural features such as the eye, the eyewall, the rainbands, and mesovortices was crucial to the domain. In addition, studying the evolutionary process of the hurricane as it starts from a disorganized cluster of clouds and intensifies to a category 5 hurricane was crucial to the overall understanding of the phenomenon.

The hurricane visualizations presented here are based on model simulations of hurricane Bonnie (August 1998), hurricane Isabel (September 2003), and hurricane Katrina (August 2005). The data for hurricane Bonnie was obtained from a MM5 model simulation, with 2 km spatial resolution and 3 minute temporal resolution for 6 hours (Braun *et al.* 2006). The data for hurricane Isabel was generated with a 2 km resolution and 1 hour temporal resolution for 48 hours with 100 levels. It was produced by the Weather Research and Forecast (WRF) model, courtesy of NCAR, and the U.S. National Science Foundation (NSF), and was the same data used for the IEEE Visualization 2004 contest (Kuo *et al.* 2004). The hurricane Katrina simulation was obtained by running the National Center for Environmental Prediction's Eta numerical weather prediction model, run at 10 km horizontal resolution and a 1 hour temporal resolution for 72 hours, with 60 vertical levels. The variables contained in our dataset are Cloud water [kg/kg], Geopotential height [gpm], Mean sea level pressure (Eta model) [Pa], Relative humidity [%], Temperature [K], u wind [m/s], v wind [m/s] and Pressure vertical velocity [Pa/s].

The current tools that application domain experts use are time-tested and effective but limited due to their two-dimensional nature (COLA 1988). Two-dimensional visualization was effective in visualizing contour plots of a single level, but exploring the three dimensional structure of the hurricane was not possible. Similarly, the change in value of an attribute from one timestep to the next or over an interval could not be directly observed.

6.3 Approach

Current hurricane visualization tools facilitate the exploration of dynamic structures using a combination of isosurfacing, volume rendering, and vector visualization to visualize the attributes in the hurricane. These techniques can fail to convey internal hurricane structures, especially in the eyewall region. The hurricane structures around the eye contain crucial information that can be used to predict changes in the intensity of the hurricane. When a simultaneous change in humidity and vertical velocity over a small number of timesteps is observed, it is often followed by a change in intensity of the hurricane. Visualizing such change could be important for the prediction of hurricane intensification.

Visualizing hurricanes in an illustrative style provides crucial morphological and temporal information that is not always apparent from standard visualization techniques. We apply illustration-based techniques such as boundary enhancement and silhouette enhancement to accentuate internal features in the underlying data. We found that the silhouette enhancement algorithm also helps in the identification of vertical wind shear in a hurricane. Illustrative visualization was useful in investigating vortex rollup and mesovortex formations. Such phenomena are believed to be linked to intensity change and so visualizations that allow exploration of these phenomena are very important. We convey temporal change in the data by applying our illustration-inspired techniques, introduced in Chapter 4: specifically speedlines, flow ribbons, strobe silhouettes, and opacity-based techniques.

6.4 Understanding spatial distributions

Visualizing the morphology of a single timestep is critical to the overall understanding of the hurricane. A visual representation of each timestep can be obtained by visualizing the various scalar quantities such as cloud water, humidity, and temperature. Currently used tools provide a two-dimensional visualization of an attribute such as humidity at a certain level (height) from the sea level, as shown in Figure 6.2. The three-dimensional structure of the hurricane cannot be visualized and understood using this method. We generated a volume rendered representation of the hurricane to provide a three-dimensional representation of the hurricane.

Traditional raycasting-based visualization tends to limit the visualization of internal structures of the hurricane, due to the obscuration of voxels farther from the eye by voxels closer to the eye. Volume illustration techniques were developed by Rheingans and Ebert (Rheingans and Ebert 2001) that were some of the first techniques that were inspired by illustrations. They identified features in volumetric data using gradients and accentuated key features, while de-emphasizing uninteresting regions (Rheingans and Ebert 2001). These techniques generated expressive visualizations of scientific data by using data dependent characteristics such as the gradient magnitude and direction. One of the techniques they identified locates the boundaries between features in the dataset by identifying regions of large gradients, then accentuates these identified boundaries. Another technique, they developed, identifies and accentuates the silhouette of the structural features in the data. This techniques helps to identify internal features such as the eyewall, surrounding rainbands, and other features that are significant for exploratory and investigative purposes.

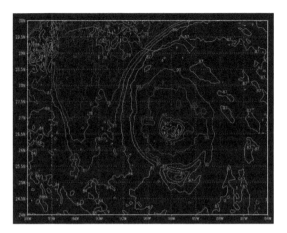

FIG. 6.2. This figure shows a two-dimensional visualization of humidity in a single level in hurricane Katrina being visualized using isocontours.

The humidity attribute is particularly useful for domain experts to visualize the evolution and transformation of the storm. Figure 6.3 shows a set of examples in which our techniques generated more insightful images than standard visualization techniques. The left image depicts a standard volume rendered image of the humidity attribute with lighting. The eye is located towards the top left of the image. The middle image shows a visualization in which the boundary enhancement technique has been applied to enhance feature boundaries, namely those of the rainbands and the eyewall. The right image is of a visualization generated using the silhouette enhancement technique in which the eyewall and the rainbands are accentuated while the unimportant details are removed. The width of the eyewall and the structure of the rainbands can be seen most clearly in this visualization.

6.4.1 Investigating vertical wind shear

Visualizing the hurricane using knowledge about the intensification and dissipation of hurricanes produces more effective and informative visualizations. It is known that the presence of *vertical wind shear* leads to hurricane weakening and dissipation. Vertical wind shear is the dif-

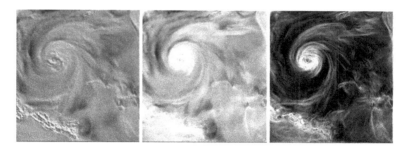

FIG. 6.3. The left image depicts a standard volume rendered image of humidity for timestep 71 for hurricane Katrina. The central image depicts the boundary enhanced version of the same image with the eye, eyewall and the rainbands being seen. The right image depicts the silhouette enhanced version with clearer depiction of the features.

ference in the wind speed and direction between the upper and lower atmosphere. If there is large vertical shear the hurricane cannot form and hold itself vertically which eventually leads to its dissipation.

To identify such vertical wind shear, a top-down view to compute silhouettes is most suitable. The silhouette computation considers voxels whose gradient is perpendicular to the view vector. This is ideal for identifying vertical wind shear: a strong, crisp, well defined silhouette implies less vertical wind shear and a blurred, diffused silhouette indicates the presence of a large vertical wind shear because the structure is more vertical coherent. In this manner, not only can the presence of vertical wind shear be identified, but it can be quantified by the strength and crispness of the observed silhouette. The equation to produce this silhouette enhancement for the investigation of vertical wind shear is

$$o_s = o_v(k_{sc} + k_{ss}(1 - abs(\nabla f_n \cdot V))^{k_{se}})$$

where o_s is final silhouette enhanced opacity of the voxel, o_v is the original opacity of the voxel, ∇f_n is the gradient of the current volume sample and V is the view vector, k_{sc} controls the scaling of the non-silhouette regions, k_{ss} controls the amount of silhouette enhancement, and k_{se} controls the sharpness of the silhouette curve.

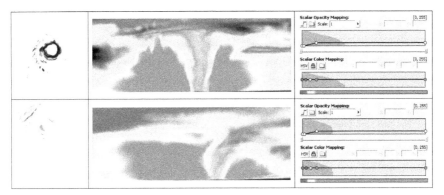

FIG. 6.4. This image depicts the silhouette of the wind speeds and a cross-sectional view of the same time step. The top row shows a completely formed circular silhouette that indicates no vertical wind shear, as can be verified from the right cross-sectional view. The bottom row shows a light, barely visible silhouette that indicates high vertical wind shear. The cross sectional view confirms the presence of high vertical wind shear in the hurricane at that timestep.

In order to validate our vertical wind shear identification, we generated visualizations to investigate the shear in a particular timestep based on its silhouette-enhanced image. Figure 6.4 shows the results. The left column depicts the silhouette-enhanced image for timesteps 7 and 46. The middle column shows a cross-sectional view of the wind speeds volume and the lower image shows the vertical shear towards the right in the hurricane. The right column contains the transfer functions used to generate the cross sectional visualizations. The top image in the left column shows a strong, well defined silhouette that correlates with the strong vertical coherence of the hurricane that can be seen in the top image in the middle column. On the other hand, the lower silhouette image is blurred and can barely be seen. This correlates with the large amount of rightward vertical wind shear that can be seen in the lower image in the middle column.

6.4.2 Investigating the structure of vorticity waves in the eyewall

Theoretical studies have shown that a rapidly spinning column of fluid will become unstable, causing a wavelike perturbations to grow. In Figure 6.5, which is an illustrative visualization of cloud water in hurricane Katrina, we can see this type of instability showing a nearly vertically aligned wave with an altitude-dependent amplitude in the cloud water field. Due to this instability, some cloud water can break off from the main eyewall. This image shows a depiction of the hurricane eyewall just before the hook-shaped cloud breaks off from the hurricane. These instabilities in the eye of the hurricane are clearly associated with regions of strong upward motion (upwelling). Rapidly rising air stretches in the vertical direction and vorticity is spun up by vortex tube stretching. The basic principle is conservation of angular momentum, analogous to a spinning skater who draws in her arms to increase her speed. To investigate this instability further, we visualized the vorticity attribute in conjunction with the upward (vertical) velocity.

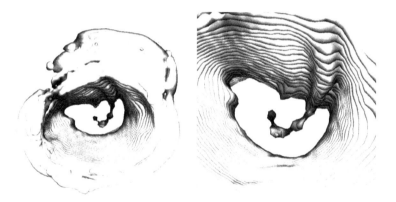

FIG. 6.5. The left image was generated using the boundary enhancement technique on the cloud liquid water attribute of the 53rd timestep. The right image shows a closer look at the hook in an illustrative style.

The top image in Figure 6.6 is a volume rendered image of the vorticity and vertical velocity fields of the first time step of hurricane Bonnie. In this image, the vertical wind velocity is rendered in red and the positive vorticity is rendered in green. Dark red regions are regions of strong positive vertical velocity; green represents regions where the positive (counterclockwise) vorticity is large. The technique emphasizes regions that are vertically aligned. The outer band of green in the northern part of the eyewall coincides with strong upwelling. The vorticity changes sign in the upper part of the storm; therefore the restriction to positive vorticity enables us to focus on the

FIG. 6.6. The top image is a volume rendered image of the vorticity, rendered in green, and vertical velocity, rendered in red, of the first time step of hurricane Bonnie. Dark red regions are regions of strong positive vertical velocity, bright green regions are high positive vorticity regions. The bottom image depicts an illustration-style rendered image of the same positive vorticity. It provides a three-dimensional representation that clarifies that the double green rings visible in the left image are actually observed at different levels in the data.

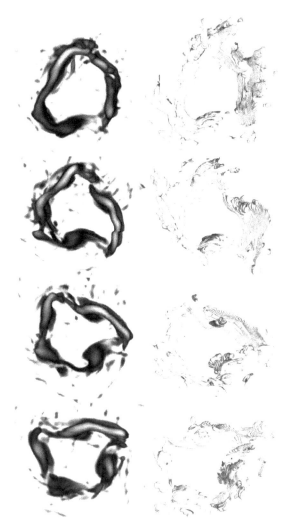

FIG. 6.7. Time sequence showing various stages of vortex rollup. The left column shows a series of volume rendered snapshots of the 13th, 26th, 30th, and 33rd timestep. The right column shows a illustration-style visualization of the corresponding timesteps. The illustrative images greatly help in an increased understanding of the three-dimensional nature of the vortex rollup.

lower portion of the storm where release of latent heat leads to the rising of warm, buoyant air. The bottom image in Figure 6.6 is an illustration-style rendering of the same positive vorticity. The illustrative visualization shows that the two green vorticity rings in the top image are actually observed at different levels and can be seen as a ledge.

The top image in Figure 6.6 also shows localized patches of high vorticity (green - does not show closed circulations) around the outer vorticity ring. Those vorticity regions are co-located with high vertical velocity regions, which links strong upward convection (upwelling) in regions of high vorticity. One explanation for this is that, rapidly rising air (signified by high vertical velocity) stretches in the vertical direction and vorticity is spun up by vortex tube stretching. This process is similar to the flowing of water from a bathtub into a drain.

The illustration-inspired image provides depth cues that a volume rendered image with lighting cannot provide. The illustration-stylized image facilitates the visualization of mesovortices, allowing the viewer to visualize the varied depths at which these vortex structures occur. Such illustrative images help in investigating and understanding the interactions between the vorticity at different levels.

A spinning ring of fluid is known to become unstable when the radial wind shear becomes sufficiently large (Drazin and Reid 1981). In this case, the vortex ring rolls up into localized vortices, as can be seen clearly in Figure 6.8. This process again leads to mesovortices, but unlike the vortex stretching mechanism discussed earlier, it does not necessarily involve vertical motion. Figure 6.7 is a time sequence showing various stages of a vortex rollup. The left column shows volume rendered images of the vorticity in a top-down view. The images in the left column are the 13th, 26th, 30th, and 33rd timestep of the vorticity respectively.

This view allows us to look more carefully at the characteristics of mesovortex formation in the lowest part of the storm. The first image in the left column shows that the vortex has developed a wavelike perturbation, which subsequently increases in amplitude. Vortex rollup begins in the subsequent image and continues in the third image from the top. In the last image in the left column, the eyewall vorticity has become concentrated into four or five mesovortices. Figure 6.9 shows that the mesovortices formed near the bottom of the hurricane are not correlated with strong

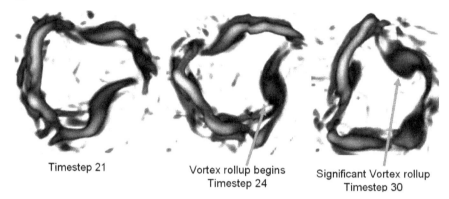

Timestep 21   Vortex rollup begins   Significant Vortex rollup
                        Timestep 24   Timestep 30

FIG. 6.8. These images show snapshots from three different timesteps of vorticity in hurricane Bonnie. They show the vortex rollup phenomenon clearly.

vertical velocity.

The illustration style of the same sequence, shown in the right column in Figure 6.7, gives a clearer view of the complex 3D nature of vortex rollup. This view shows the vertical alignment of the eyewall mesovortices, and shows the intricate nature of the vortex dynamics occurring in the eyewall. The illustrative style helps the viewer to understand that the vortex rollup is a three-dimensional phenomenon and does not occur only at the lower level.

6.5   Understanding and exploring the time-varying nature of a hurricane

Visualizing a snapshot of each timestep individually limits a user's ability to understand the trends and transformations that the hurricane undergoes. For such time-varying datasets, visualizing volume rendered snapshots of each timestep or viewing an animation of rendered images of each timestep may be of limited use. These snapshots or animations may not convey change in the structure and position of the contained features in the time-varying dataset.

We have used illustration-based principles to visualize the change undergone by the hurricane. Using these principles, we can convey morphological and temporal changes in the hurricane using a single image. Opacity-based techniques in combination with speedlines and strobe silhouettes,

FIG. 6.9. This image shows the positive vorticity rendered in green and vertical velocity rendered in green for the 33rd time step, corresponding to the last row in Figure 6.7. The image shows that the mesovortices formed near the bottom of the hurricane are not correlated with strong vertical velocity: that is, there is almost no vertical velocity in the strong positive vorticity regions shown in green.

convey change over time effectively.

6.5.1 Investigation of the motion of the hurricane center

The translatory motion of the hurricane is accompanied by rotation around its axis. The direction of motion of the hurricane center is critical to the understanding of the "wobbling" motion of the hurricane as it progresses along its path. The wobbling motion of the hurricane's axis of rotation, also known as precession, may be linked to stability due to shear in the upper levels.

Illustrators use trailing lines to create the effect of motion in still images (McCloud 1994). Upon observing the properties of these lines, we found that illustrators encode temporal information in them. Our visual system processes these trailing lines as evidence of the object having been there in the past and having moved to its current state. We call these trailing lines *speedlines* and have previously shown their use in visualizing time-varying CFD data (Joshi and Rheingans 2005). We applied the speedlines technique to visualize hurricane data. Figure 6.10 depicts a visualization of the hurricane with speedlines. The eye of the hurricane is tracked automatically by identifying the lowest mean sea level pressure and the minimum wind speeds in every timestep. This gives an idea of the track the hurricane has followed. The path is then offset in a direction perpendicular to

100

FIG. 6.10. This image depicts a pen-and-ink style rendered visualization of timestep 44 of the hurricane. The annotation of the image with speedlines provides temporal context and conveys information regarding the upwards curving motion of the hurricane.

the motion in each direction to obtain speedlines. The hurricane itself is visualized in a pen-and-ink shaded style to be consistent with the illustrative paradigm of the image. The annotation of the image with speedlines provides the effect of motion and conveys important information regarding the upwards curving motion of the hurricane. The speedlines also show a cycloid motion of the hurricane center due to an overall rotation about the center. This may be a useful way to depict "wave 1" Rossby modes, which involve a precession of the top of the storm relative to the lower levels.

6.5.2   Examining the rate of intensification

The speed at which a hurricane intensifies is critical in the overall understanding of the intensification process. For example, an intensification of a category 2 hurricane to a category 5 in a few hours could prompt a further investigation into the conditions that caused such rapid intensification. Studying the rate of intensification is also crucial in correlating the trends in observed values of attributes of the hurricane.

This problem was approached by using a technique that has been inspired by illustrators who

FIG. 6.11. This image depicts the use of opacity-based techniques to convey the transformation of category 1 hurricane into a category 5 hurricane. The transformation from the topmost faded rendition of the hurricane at timestep 10 to the bottommost rendition is at timestep 71 captures the temporal nature of the evolution. The image shows a visualization of hurricane Katrina over a 72-hour period.

use blurring and desaturation for older timesteps and bright, crisp renditions of newer timesteps (MacCurdy 1954), (McCloud 1994). Such illustrations are generally used to convey past positions and structure of the features in the current timestep. This technique provides context to the visualization of the current timestep by providing faded visualizations of older timesteps that convey change in shape, orientation, and position. To generate a visualization using this technique, certain interesting timesteps are merged into a single timestep to create a composite volume. The visualization is then created, taking into consideration the age of each component timestep. Visualizing change over time in a single image can be achieved using this technique.

Figure 6.11 shows a visualization of the current timestep with the older timesteps fading out. The oldest timestep (timestep 10) is visible and shows a faded representation of the hurricane at category 1. The subsequent timesteps (timestep 34 and 53) are when the hurricane is more symmetric with an eye and an eyewall. By timestep 71, the hurricane is a full-blown category 5

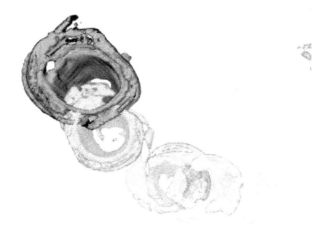

FIG. 6.12. This image depicts the use of opacity-based techniques to convey the intensification of hurricane Katrina. The visualization contains a footprint for every category change it undergoes. The first footprint, on the bottom left of the image, is the change to category 2. The hurricane intensifies very quickly after that, as per the subsequent two footprints and then the last footprint signifies the intensification to category 5.

hurricane. The evolution of a hurricane from a small organized cluster to a category 5 hurricane is conveyed from this visualization.

Figure 6.12 contains footprints of the timesteps when the hurricane changes its category. This produces a visualization that depicts a snapshot of the hurricane at every step in the intensification process. This image depicts the intensification of the hurricane from a category 2 onwards. The last footprint is for the hurricane when it intensifies to a category 5 hurricane. The rapid intensification of the hurricane is visible from the image. The hurricane is a category 2 hurricane initially and quickly intensifies in a matter of a few hours to a category 4, after which it intensifies to a category 5, as indicated by the last footprint.

Some of these techniques can be used in combination to produce more informative visualizations. For example, Figure 6.13 shows the opacity-based technique in conjunction with speedlines to visualize a combination of the 10th, 35th, 53rd, and 71st timesteps. The change in size, structure

FIG. 6.13. The image depicts the combination of opacity-based techniques and speedlines to convey the motion of hurricane Katrina. The hurricane moves upwards and leftwards. The thickness of the speedlines reduces as it approaches the current timestep. The image depicts the visualization of time steps 10, 35, 53 and 71 of the cloud water quantity. The growth in size of the hurricane and the internal structure of the hurricane are visible.

and position of the hurricane is conveyed from the image. The aim of the image is to capture the upward and leftward motion of the storm. The structural changes in the hurricane as it intensifies over each 24-hour period can be seen by the increasing size of the hurricane. The speedlines technique provides additional temporal cues to convey change in time. The opacity-based techniques can also be used to draw the user's attention to a specific timestep of interest.

6.5.3 Investigating the temporal and structural change in size of the hurricane

Visualizing past positions of the storm can be useful in conveying the direction in which the hurricane is progressing. Illustrators have used trailing silhouettes to depict past positions as well as the direction of motion of an object over time (McCloud 1994), (Kawagishi *et al.* 2003). Trailing silhouettes have been effectively used to convey past positions and the general direction of the feature in a time-varying dataset (Joshi and Rheingans 2005).

We use the strobe silhouettes technique to depict the direction of motion as well as past po-

FIG. 6.14. This image depicts the strobe silhouettes technique being applied to the hurricane dataset. The cloud water quantity for timestep 71 is visualized in an illustrative style with trailing silhouettes for timestep 10, 35 and 53 being depicted in the image.

sitions of the hurricane in a single image without cluttering the visualization. The direction-of-motion vector is calculated based on the change in position of the eye of the hurricane and is used to compute the strobe silhouettes.

Figure 6.14 shows the past positions of three such timesteps, along with an illustrative rendition of the final (71st) timestep of the cloud water quantity. The past positions of timesteps 10, 35, and 53 are depicted by trailing silhouettes. The strobe silhouettes technique is particularly effective at conveying the past positions of the hurricane while also visualizing the current timestep in full detail. This provides temporal context and conveys information regarding the change in size of the storm. The downward direction, increase in size of the hurricane, and origin of the storm are conveyed effectively in Figure 6.14 using strobe silhouettes.

6.6   Expert Evaluation

I evaluated the illustration-inspired techniques with six expert users from the application domain who gave subjective feedback regarding the usability of my techniques. As has been mentioned by Kosara *et al.* (2003), "the input of application domain experts is worth its weight in
gold." I wanted to evaluate the benefit of my techniques in the application domain of hurricane
visualization, which was my primary domain of focus.

Our users were

- Scott Braun, Research Meteorologist at NASA's Goddard Space Flight Center.

- Julio Bachmeister, Associate Research Scientist at the Goddard Earth Sciences and Technology Center (GEST).

- Bill Olson, Research Associate Professor at the Joint Center for Earth Systems Technology.

- Lynn Sparling, Associate Professor of Physics at UMBC and a UMBC-affiliated member of
  JCET.

- Hai Zhang, Postdoctoral Researcher in the Physics Department at UMBC

- Samuel Trahan, Graduate Student in the Physics Department at UMBC.

For this purpose, we created a website containing images generated from visualization pack-
ages that the experts use on a regular basis, such as GrADS (COLA 1988), as well as our
illustration-inspired visualizations. The website is at

`http://www.cs.umbc.edu/~alark1/hurricane_study.php`.

Figure 6.15 shows a screenshot of the first page of the website of the expert evaluation.

6.6.1   Procedure

The procedure for this expert evaluation was:

106

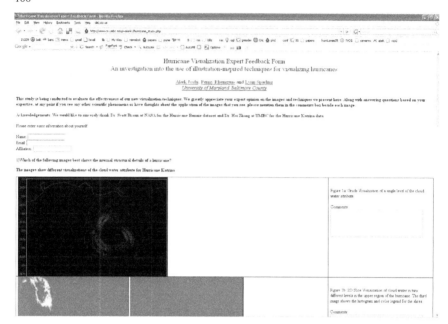

FIG. 6.15. A screenshot of the first page of the online expert evaluation of our techniques.

1. Explain the purpose of the expert evaluation to the domain experts.

2. Get their information for the expert evaluation.

3. Conduct the expert evaluation.

6.6.2 Datasets

For this expert evaluation, two hurricane datasets were used. The first dataset was a simulation of hurricane Katrina which was provided by Dr. Hai Zhang, Dr. Miodrag Rancic, and Dr. Lynn Sparling at UMBC. The dataset is $276 \times 288 \times 60$ in size with 72 timesteps simulated at one-hour intervals. Each timestep includes values for total precipitation, cloud water, geopotential height, mean sea level pressure, relative humidity, temperature, u wind vectors, v wind vectors and pressure vertical velocity.

The second dataset was a simulation of hurricane Bonnie which was obtained from Dr. Scott Braun at NASA. The dataset is $250 \times 250 \times 27$ in size with 120 timesteps at three-minute intervals. Each timestep included values for radar reflectivity, vertical velocity, temperature, potential temperature, divergence, u wind vector, v wind vector, vorticity, and pressure.

6.6.3 Expert feedback

We found that the experts preferred our illustration-inspired techniques over standard visualization techniques in most cases. The experts found our visualization techniques to be very useful for examining hurricane structure, three-dimensional mesovortex formation, and visualization of the current timestep with context. For example, Figure 6.5 was preferred over 2D isocontour (Figure 6.18) and the 3D volume rendering images (Figure 6.16) by all of the experts for its ability to clearly visualize the internal structure of the hurricanes. One expert says, *"This is an exceptionally effective portrayal of the cloud water field. Here we can see a wave perturbation in the eyewall that has an altitude dependent amplitude. This type of visualization has a lot of potential for investigating the dynamics in the hurricane inner core."* The illustrative visualization in Figure 6.5 was found to be more suited than standard visualizations for answering questions regarding the internal structure of a single timestep in a hurricane. Figure 6.18 was useful for answering other hurricane related questions, which we discuss later.

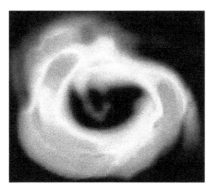

FIG. 6.16. This image shows a volume rendering of the cloud water quantity in hurricane Katrina.

FIG. 6.17. This image shows four snapshots of cloud water being rendered in hurricane Katrina at four different timesteps. The direction of motion is not easily apparent by just looking at this set of images.

Experts preferred the illustrative pen-and-ink style visualizations in the right column to the standard volume rendered vorticity images in the left column in Figure 6.7. The ability to clearly visualize the three dimensionality of the hurricane in the illustration-inspired visualizations helped understand the structural details of the hurricane. One expert writes, *"The illustration-based technique on the right is radically different from anything currently used in hurricane research. The cases here show vortex rollup in 3D. This is a very promising new way to look at hurricane inner core structures and dynamics."* Using the illustrative visualization, the hurricane experts were able to answer questions regarding the presence and structure of mesovortices and to further investigate the vortex rollup phenomenon.

The simple illustrative techniques, such as the boundary and silhouette enhancements shown in Figure 6.3 were preferred over standard volume rendering techniques. The ability of the illustrative images to clearly show features such as the eyewall and rainbands was appealing to them. Based on Figure 6.4, one expert notes the value in using silhouette enhancement and writes, *"This is very useful for looking at the effects of vertical wind shear on the hurricane structure."* This technique helped hurricane experts to answer questions regarding the effect of vertical wind shear on the dissipation and eventual disintegration of a hurricane.

Figure 6.12 was liked by all six experts one of whom wrote, *"This is a great way to visually track the motion of the storm."* Another expert writes, *"The figure might be very useful in a study of category change in response to topography etc. during landfall, wind shear or sea surface temperatures."* Figure 6.11 and 6.13 were both preferred to looking at snapshots of each timestep individually or viewing the panorama of constituent timesteps shown in Figure 6.17. They liked the fact that both the storm motion and the structural evolution could be clearly seen in the illustration-inspired images. Figure 6.10 was found to be effective and the illustrative style found to be novel. The visualizations helped them answer questions regarding the direction of motion of the hurricane, the rate of intensification of the hurricane, and the evolution of the hurricane at various stages along its track.

For Figure 6.6, an expert wrote, *"This is an effective way of showing the correlation; this particular plot shows that a region of strong, vertically coherent vorticity marks the inner region of upwelling. The problem here is that we cannot tell if the green annular features are displaced in the vertical or if we are seeing secondary eyewall formation. The illustrative visualization helps clarify that by showing that the features are accruing at two different vertical levels."* The visualization helped experts correlate the vorticity and the vertical velocity variables in the data and led them to notice that the two vorticity rings in the hurricane occur at different altitudes.

In some cases, the traditional 2D visualization techniques were preferred. We suspect that this might be due to the familiarity with a particular tool or due to their training in using a particular 2D visualization technique. In particular, the integration concept of volume rendering was confusing to the experts, since their training in interpreting 2D visualizations led them to interpret the values as being that of the topmost level. Experts prefer to investigate the values at particular levels separately, so visualizing a composite image seemed unintuitive to them. Despite the fact that volume rendering images were less preferred, our illustrative representations such as the boundary and silhouette-enhanced visualizations, shown in Figure 6.3, were liked due to the fact that they enabled them to look at three-dimensional features such as the rainbands, the structure of which is not as clearly visible in 2D visualizations. Figure 6.18 shows an example of a 2D visualization technique (isocontours) being used to visualize a single level in the hurricane. An expert mentioned that such an image can reveal the 2D structure in the eyewall clearly but the 3D structure cannot be seen.

Figure 6.19 shows another example of wind vector visualization using hedgehogs (small ar-rows indicating wind direction and speed) in a single level. The preferred visualizing the wind

FIG. 6.18. This image depicts a 2D visualization showing isocontours for cloud water in hurricane Katrina at a single level.

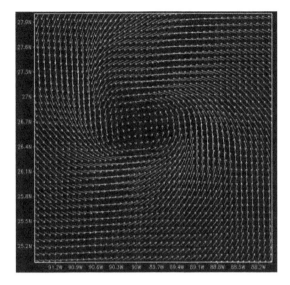

FIG. 6.19. This image depicts a 2D visualization showing "hedgehogs" for visualizing wind vectors in hurricane Katrina at a single level.

quantity in this manner than visualizing just the windspeeds in three-dimensions. Visualizing wind vectors in this manner, even a single two-dimensional level, helped them visualize both the magnitude and direction of the winds. They like the fact that it showed the wind structure at a single level nicely. In such cases, we believe that the texture-advection techniques introduced by Schafhitzel et al. (2004) could easily be adapted for three-dimensional visualization of wind vectors.

Chapter 7

POINTILLISM-BASED VISUALIZATION OF ATTRIBUTE CHANGE IN TIME-VARYING DATA

Standard visualization techniques for time-varying data (such as snapshots or animations) are limited at conveying temporal information to the viewer. Time-varying data invariably contains multiple attributes, such as temperature, rainfall and so on, at each location in space. The complexity of viewing and understanding the data is increased by the fact that these attributes are constantly changing over time. A domain scientist would like to get a better understanding of the nature of change that a particular attribute is undergoing over time. Our discussions with domain experts led us to a realization that they need to understand the change in attribute value in a particular region in their data. For example, in our application domain of hurricanes our collaborators would like to understand the changes a hurricane is undergoing as it intensifies. Various attributes in a hurricane such as humidity and windspeed are of interest to them and specifically particular regions in the hurricane that are undergoing change in values. Existing approaches either produce snapshots of each timestep or an animation requiring them to be able to understand the changes the hurricane is undergoing over time.

Alternatively, the change can be visualized by computing the difference in values and then visualizing the computed difference. The limitations of these approaches is that they are not able to capture the intermediate values and provide a comprehensive unifying picture that conveys attribute change over the time interval.

I developed a novel visualization technique that conveys the attribute change in a single image

112

as well as convey intermediate values in the specified time interval to allow the user to visualize trends in the data.

Pointillism is a technique in painterly rendering where the artist paints by placing brush strokes in the form of points on the canvas. The brush strokes form a cohesive picture when looked at from a distance, but on looking closely, they appear to be a seemingly incoherent collection of brush strokes. This painting style was pioneered by Seurat and Signac around the 1890s. The most famous examples of pointillist painting are Seurat's "Sunday Afternoon on the Island of La Grande Jatte" and "La Parade de Cirque" (Figure 7.1). The paintings look beautiful from a distance, but on close inspection they reveal an intricate composition of brush strokes, as can be seen in Figure 7.2. Seurat incorporated color theory and color mixing, as proposed by Rood (1879), into his paintings (Kemp 1990). Instead of the common practice of physically mixing colors on a palette, Seurat applied the color mixing theory by placing brush strokes close to each other to facilitate a visual "mixing of colors." It can be clearly seen in the closeup of the man in Figure 7.2, in which warm-colored brush strokes are placed over the underpainting of the man's head. The effect on looking at the entire painting is of a soft light being reflected onto the man's head. The ability of the human eye to visually mix the colors to obtain a certain effect is at the center of my approach to and we use it for the purpose of visualizing attribute change in time varying data.

In our approach, we use the pointillism paradigm to convey visual change in attribute values over time. Instead of computing the change over an interval of time, we sample each timestep in the interval and visually represent the data using a pointillistic painterly style. We use various attributes of painterly rendering such as brush size, brush color, and saturation to convey change visually. This provides a visual representation of the attribute distribution in space and time. In the subsequent timesteps, brush strokes are placed on the canvas in a non-overlapping manner, to mimic pointillistic painting.

In the visualization literature, Urness et al. (2003) have proved the effectiveness of color blending techniques in their "visual blending" paradigm for visualizing flow.

We demonstrate our results with synthetic data and real world data, including global rainfall, infant mortality (CDC 2002), US presidential elections per county (PurpleAmerica 2004), and

FIG. 7.1. "La Parade de Cirque", Georges Seurat, 1888. Note that the painting is made of multiple small closely placed brush strokes.

hurricane simulations over a time interval. The results clearly show the efficacy of my techniques in conveying these changes and visualizing interesting patterns using the pointillism style of painting. I conducted a user study to evaluate the effectiveness of our techniques. The user evaluation results showed that users were able to conduct tasks faster and more accurately with our technique than with traditional snapshots or animation-based techniques.

7.1 Approach

We create the pointillistic visualization over the specified interval by sampling intermediate timesteps. We use the contribution of the values in each sampled timestep to paint the canvas with brush strokes. The current timestep being visualized is given highest priority and is rendered using the largest brush size. An older timestep is visualized with a smaller brush stroke and a desaturated brush color.

We consider a simple synthetic dataset to explain our technique clearly. Figure 7.3 shows five images where the color at each images represents the value of an attribute as it changes over time.

FIG. 7.2. Detail of "La Parade de Cirque", Georges Seurat, 1888 clearly depicts the detail and arrangement of strokes. For example, in the background as well as on his suit there are brush strokes that almost seem random to the viewer at this level of detail, but on looking at Figure 7.1, it visually blends to produce a visually appealing painting.

The images convey an increasing trend in the data over time (timestep 1 to timestep 5).

7.1.1 Sampling

The sampling process described above ensures that intermediate values over the time interval are visualized. Regions of homogeneity, regions of constant increase and regions of constant decrease now become visually apparent, allowing an application domain expert to quickly identify regions that they would like to investigate further. Additionally, a viewer can now easily identify patterns such as an increase followed by a decrease or decrease followed by an increase in attribute values. Figure 7.4 shows a depiction of the sampling strategy that we use to visualize the data.

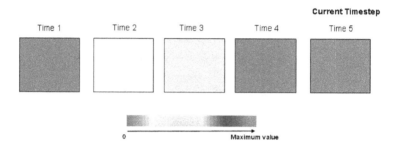

FIG. 7.3. This image depicts a simple synthetic dataset where the five images represent the values of an attribute that change over time. The increasing trend is apparent by visualizing the data in conjunction with the color scale.

7.1.2 Brush selection

Pointillistic brush strokes are not perfect circles. Previously, the pointillistic painting style has been represented in the form of circular or point sized brush strokes (Hertzmann 1998), (Saunders *et al.* 2005). We experimented with using circular brush strokes, but found the results to be too regular, as can be seen in the left image in Figure 7.7. Closely observing pointillistic paintings by Seurat, such as Figure 7.1, validate our observation that brush strokes are not perfect circular points.

In order to provide a painterly feel to the created visualization, we painted pointillist-style brush strokes on a canvas. These point brush strokes were then scanned in and used as alpha textures for the strokes being rendered on the screen. Figure 7.6 shows the sixteen brush strokes that we painted, scanned, and used in our painterly rendering process. The right image in Figure 7.7 shows a painterly visualization that looks much more organic and has a true pointillistic feel to it.

In order to simulate a painterly rendering style, we use a uniformly distributed probability density function to pick one of the sixteen textures as we place the brush stroke on the canvas.

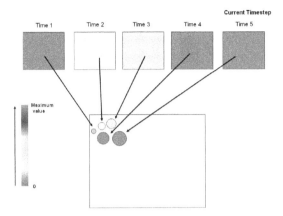

FIG. 7.4. This schematic describes our brush selection, color selection and brush placement approach. The brush size decreases as the temporal distance of the selected timestep from the current timestep increases. As can be seen in the figure, the red brush stroke is the smallest in size since its timestep (time = 1) is farthest from the current timestep (time = 5). The color of the brush stroke is selected based on a clamped rainbow color scale. The saturation of the stroke color is decreased depending on the temporal distance of the selected timestep from the current timestep. Since the first timestep (time = 1) is farthest from the current timestep (time=5) being visualized, the red brush stroke color is much more desaturated/faded. The stroke placement algorithm randomly identifies a location on the canvas that is not occupied, places a stroke there and then marks that location as occupied.

7.1.3 Color selection

The color selection process for each brush stroke was based on the attribute value as well the age or the temporal distance of the current timestep from the first timestep in the interval.

One of the main aims of our painterly visualization technique was to allow the viewer to be able to visualize the current timestep data with values from older timestep. To ensure that the older timesteps do not dominate the visualization, we reduce the saturation of the identified color by a factor that is inversely proportional to the temporal distance from the current timestep. Instead of using the complete rainbow color scale, we use a truncated rainbow color map to avoid misinterpretation due to a wrap around. We used the rainbow color map, since it can represent a large spectrum of values in the visualization. We are aware of the limitations of the rainbow color

118

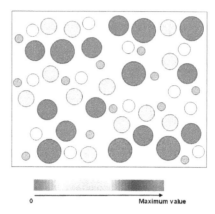

0   Maximum value

FIG. 7.5. This image shows a pointillistic visualization of the synthetic example described in Figure 7.3 and Figure 7.4. The brush strokes representing the current timestep are biggest and brightest than the brush strokes representing older timesteps. The clamped rainbow color scale shown here helps the viewer understand the increasing trend in the time-varying data. We later introduce a better color legend that allows the viewer to more easily visualize patterns in the data.

map (Rogowitz and Treinish 1998) and are investigating other more suitable colormaps for our painterly visualization.

We use the Hue-Saturation-Value (HSV) color space to pick a color for our brush stroke. The truncated rainbow color scale is used with the maximum hue value clamped at 300. The value at a particular point in the 2D space is used to look up the brush color in the HSV color space.

Saturation is given by:

$$Saturation = \frac{1.0}{d_k}$$

where d_k stands for the temporal distance of the current timestep from the newest timestep.

We again refer to Figure 7.4, in which older timesteps are represented with a more faded, desaturated brush stroke while newer timesteps are represented with brighter brush strokes. The final visualization, shown in Figure 7.5 shows the complete visualization with painterly brush strokes. The faded brush color as well as brush stroke size helps the viewer differentiate brush

FIG. 7.6. The painted brush strokes used as textures for the generation of a realistic painterly visualization.

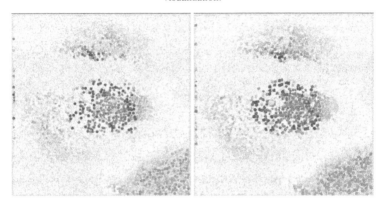

FIG. 7.7. The left image depicts a visualization in the pointillistic style with circular brush strokes. The right image shows a painterly visualization using scanned brush strokes from Figure 7.6. An organic, hand-drawn look is apparent in the right image.

strokes representing older timesteps from those representing newer timesteps.

7.1.4 Stroke placement

We observed that in pointillistic rendering the brush strokes do not overlap much (as seen in painting in Figure 7.2). We incorporated the limited overlapping of brush strokes into our system. Brush strokes are placed close to or in between brush strokes already on the canvas.

To create a pointillistic visualization, we randomly identify a location on the canvas and check whether a brush stroke is already placed there. If there is no brush stroke there, we use a uniformly distributed probability density function to pick a timestep. Based on the identified timestep, a brush

120

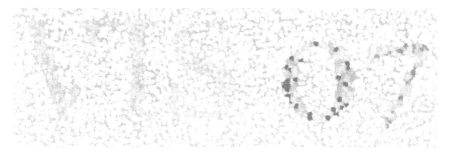

FIG. 7.8. The image depicts a synthetic time-varying dataset visualized using pointillistic tech-niques with information from five timesteps. The synthetic data was created to highlight the possi-ble variations in time-varying data. The values in the alphabet "V" are monotonically decreasing, in "I" are constant, in "S" decrease and then increase back to the same value as in the first timestep, in "0" the values increase and then decrease back to the value in the first timestep and those in "7" are monotonically increasing. These variations can be followed by using the colormap provided along with the visualization.

size and brush color is calculated. The color for the brush stroke is obtained by using the attribute value at the location in that timestep. The color selection process is used to select a color for the brush stroke. After placing a brush stroke on the canvas, we mark the region on the canvas as occupied to ensure generally non-overlapping brush strokes in the final painterly visualization. To keep track of the current state of the canvas, we maintain a canvas data structure that contains the locations and extent of each brush stroke as well as which regions of the canvas are empty. This enables the algorithm to decide whether a new brush stroke can be placed without considerable overlap with another brush stroke already on the canvas.

We show some examples in Figure 7.8 for synthetic data. This dataset was particularly cre-ated to demonstrate the strengths of our techniques. The intensity of alphabet "V" is constantly decreasing over time; the intensity of the alphabet "I" is constant over the interval; the intensity of the alphabet "S" decreases and then increases back to the intensity of the first timestep; the intensity of the number "0" increases and then decreases to the intensity of the first timestep; and the intensity of the number "7" is increasing constantly over the interval.

The alphabet "V" contains five distinct colors. The bright green maps to the initial value from the newest timestep and is depicted using the biggest brush size and the most saturated color. The other colors visible are lighter shades of green, orange, and a light shade of red. On close examination of Figure 7.8, these brush strokes can clearly be identified, mapping them to the scalar values conveys the decreasing intensity of the alphabet "V" over time. Similarly, since the values for the alphabet "I" are constant over the interval, the only color visible is green in various brush sizes and various saturated levels. On close observation of the remaining alphabets and numbers in the visualization, the decrease and subsequent increase in intensity values of "S", increase and decrease of values in "0" and constant increasing trend in "7" are clearly visible.

I experimented with using a layering-based approach where I would first paint the current timestep using the largest brush size and then fill in the "gaps" with smaller brush strokes for older timesteps. However, in such images, the painterly representation is largely dominated by the current timestep and any correlation with previous timesteps is obscured. The left image in Figure 7.9 shows an example of this effect. The first layer has the largest brush strokes and also has the most brush strokes, while the brush strokes for previous timesteps can barely be seen. There are 2033 strokes for the current timestep and 359, 234, 207, and 175 for the other subsequent timesteps. Thus, a uniformly weighted probability distribution function works better and obtains an equal contribution from each timestep in the interval. The right image in Figure 7.9 shows a uniformly distributed weighting applied to each brush size.

7.1.5   Stroke size/saturation legend

Legends are generally provided with visualizations to help the user understand the values in the visualization. To effectively understand the painterly visualization and analyze the trends in the data, using a standard colormap, as shown in Figure 7.3, is not sufficient. We generate a stroke size/saturation legend (Figure 7.10) in which the stroke sizes and saturation values used in the visualization are mapped to attribute values, assisting the user in identifying the timestep to which each stroke belongs.

In the process of interacting with the painterly visualization, the user may be interested in examining certain regions of interest in more detail. As in an actual pointillistic painting - where

122

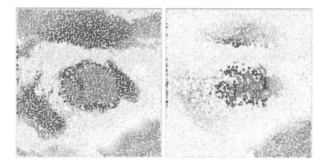

FIG. 7.9. The left image shows the effect of using a large brush stroke layer followed by placing brush strokes in gaps that remain after the first layer. In such a process, the first layer covers a large part of the painting and a very small fraction of the painting is then painted by brush strokes representing subsequent timesteps. The current timestep is represented by 2033 brush strokes, while the subsequent four timesteps are represented using 359, 234, 207, and 175 brush strokes respectively. The right image shows a painterly visualization with equal number of brush strokes per time step.

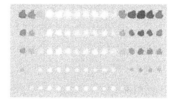

FIG. 7.10. Stroke size/saturation legend depicting the decreasing stroke size and saturation as we traverse downwards on the time axis and stroke hue values increasing as we move rightwards.

the brush strokes seem larger as one gets closer to the painting - so do our brush strokes get larger as the user zooms into the area of interest. In order to continue to provide effective visualization, our legend automatically resizes to show the larger brush strokes, as seen in Figure 7.23.

7.2   Results

We applied our techniques to synthetic data and real-world data. The synthetic data was chosen to particularly highlight the strengths of our techniques. I also show examples of our pointillism-based techniques applied to infant mortality data per state for the entire country over an eight-year period, US presidential elections results per county from 1960-2004, and hurricane data from hurricanes Katrina and Isabel.

7.2.1 Synthetic data

We show examples of the various trends that can be seen using our techniques. Figure 7.11 shows two such example trends that can be identified by using the color scale below. The left image shows an increasing trend, as can be seen by small, light red brush strokes, medium-sized yellow brush strokes, and big green and bright blue brush strokes. The right image in the figure shows a decreasing trend which can be identified by observing small light blue brush strokes, medium-sized green strokes, and larger orange and red brush strokes.

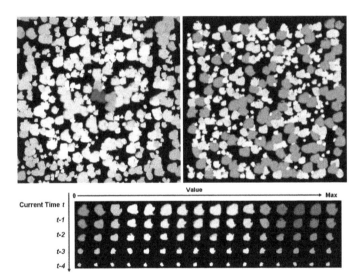

FIG. 7.11. This set of images shows some possible trends that can be clearly visualized using the scale shown below. The top left image shows an increasing trend as can be seen by the small, faded red brush strokes and large, bright blue and green strokes in the center. The top right image shows a decreasing trend where older values are represented by small, light blue brush strokes and new values are represented by big, bright red brush strokes.

124

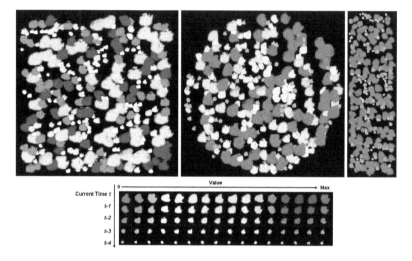

FIG. 7.12. This set of images shows some more possible trends that can be clearly visualized using the scale shown below. The left image shows an increase followed by a decrease as can be seen by small yellow strokes, medium sized violet brush strokes and big green brush stroke. The middle image in the second row shows another increase followed by decrease. The third image in the second row shows various shades and sizes of blue brush strokes which implies that the value stays constant over time.

Figure 7.12 shows examples for other trends/patterns. The left image shows an increase in values followed by a decrease. The increasing trend can be seen by the fact that the smallest brush stroke is light yellow in color, followed by light blue brush strokes and medium-sized violet brush strokes. The subsequent decrease is conveyed using a brighter, larger blue brush stroke followed by the largest, green brush stroke. In this manner, an increase-then-decrease pattern can be identified. The second image from the left shows a decrease followed by an increasing trend in a single image. The decrease is conveyed by the small violet, light blue and, light yellowish green brush strokes. The subsequent increase is conveyed by the bigger green brush stroke and the largest, bright, blue stroke. The third image shows various shades and sizes of blue brush strokes. This implies that the value does not change over time. The lack of variability conveys the constancy over time.

7.2.2 Infant mortality

We applied our techniques to infant mortality data for the United States that was collected over eight years from 1995-2002. The infant mortality data was obtained from the US Centers for Disease Control and Prevention, National Center for Health Statistics.

Figure 7.13 shows five sample snapshots of the data in time, depicting the infant mortality over a five year period from 1998-2002 based on the color scale shown.

The problem with visualizing such snapshots is that interesting trends, variability, unusual patterns and consistently minimal/maximal values cannot be easily identified. For example, in Figure 7.13, one requires considerable effort to identify the state that consistently has the maximum infant mortality rate over the five-year time interval shown here.

FIG. 7.13. This set of image depicts infant mortality in the U.S. over a period of five years from 1998-2002. The color scale provided can be used to understand the data. Identification of clear, interesting patterns and unusual behavior in the data cannot be clearly seen by looking at these snapshots. *Data credits: US Centers for Disease Control and Prevention, National Center for Health Statistics.*

In contrast, Figure 7.14 shows a pointillistic visualization that shows an overview of the data over the five-year period from 1998-2002. The visualization can help us answer questions such as which state consistently has the highest infant mortality over the interval or which state has the lowest infant mortality over the interval.

As with visualization techniques, more interesting patterns can be found by interacting with the data. As with the painting paradigm, we can get "closer" to the visualization to investigate

FIG. 7.14. This image shows a pointillistic visualization of the entire nation showing infant mortality over the time interval from 1998-2002. The visualization can clearly shows that states such as California, Utah and most of the Northeast have a very low infant mortality rate. On the other hand, states like Mississippi jump out as consistently having the highest infant mortality rate in the nation. Figure 7.15 shows a closer look at Mississippi. Other interesting patterns can be observed by zooming into a region.

patterns/trends better. Figure 7.15 shows one such example: the state of Mississippi clearly shows maximum infant mortality according to the color scale shown on the right side of the image. Alabama had high infant mortality in the past, but the value has dropped lately, as can be seen by bright, blue brush strokes in comparison to faded, violet brush strokes.

Another interesting pattern, in the form of a spike, can be observed in the state of Alaska. Figure 7.16 shows a visualization of Alaska in which we can see large greenish-blue brush strokes followed by the biggest brush stroke in red. This conveys that there was a spike in data values in 2001 (represented by the greenish blue brush strokes). The value in 2002 dropped again, as is shown by the large red brush strokes.

Figure 7.17 shows an closeup example of North Dakota, South Dakota, and Minnesota being visualized using our technique. Minnesota appears to have low infant mortality rates with some intermediate moderate values, as shown by faded yellow brush strokes. North Dakota's infant mortality seems to be falling, as can be seen by bigger yellow brush strokes, in contrast with older blue brush strokes of various shades. South Dakota, on the other hand, seems to have had higher

FIG. 7.15. This image depicts a visualization of southeastern region. The state of Mississippi's consistently high infant mortality rate over the five year period from 1998-2002 is apparent from this visualization.

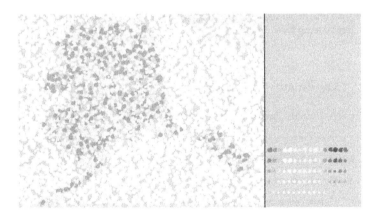

FIG. 7.16. This image depicts a visualization of Alaska in the pointillistic style. The values are generally low as can be seen by small orange and, yellow brush strokes but a spike can be clearly seen in the form of bluish green brush strokes. The values are again back to a lower value, in the most recent year, as seen in the large brush strokes.

128

values (light blue brush strokes) followed by a drop (bigger faded red brush strokes) and then an increase (even bigger green brush strokes).

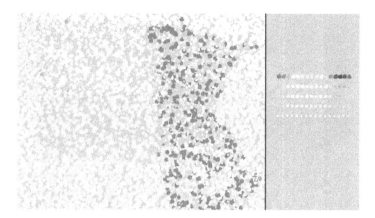

FIG. 7.17. This image depicts a visualization of the upper midwestern region. The state Minnesota has low infant mortality with some moderate values as seen by the faded yellow brush strokes. The infant mortality in North Dakota is falling as can be seen by larger, brighter yellow strokes as compared to older, light blue strokes. South Dakota has an interesting pattern, where the values were high (light blue brush strokes) and then following a dip (faded red brush strokes) are on the rise again (largest green brush strokes).

7.2.3 US Presidential election results

We applied our techniques to visualize the US Presidential election results from 1960-2004, collected at a county level. Figure 7.18 shows a snapshot of the elections from 2004 where red stands for Republican and blue stands for Democrat. Visualizing only the snapshots from a single election year precludes showing the interesting patterns in the data over time.

Figure 7.19 shows an example of our pointillistic visualization highlighting patterns in the underlying time-varying data. The top row shows three snapshots depicting results from 1988, 1992 and, 1996; the second row shows snapshots from 2000 and 2004. The pointillistic visualization clearly highlights interesting patterns, such as the fact that a section in otherwise Republican southwestern Texas consistently voted Democrat over the entire interval. There are also some regions in

FIG. 7.18. This image shows a visual representation of US voting results per county.

Colorado that consistently voted Democrat over the last five elections.

Figure 7.20 shows snapshots for the elections held in 1960, 1964, and 1968 in the top row. The bottom image shows a pointillistic visualization of the data. The variations in the southeast are clearly visible in a single image, compared to visualizing each snapshot separately. The variation can be explained by the surge in popularity of the " Dixiecrats" in the southern region. The variation is particularly high, since in 1960 the Dixiecrats won a large part of Mississippi counties and then a major resurgence in 1968 saw them winning a large part of the south, before they merged with the Republican party for the 1972 elections. The variation in the region is clearly conveyed in a single visualization using our technique.

7.2.4 Hurricane visualization

In the hurricane domain, the major challenge is the change that a hurricane undergoes within a certain time interval. Our collaborators in UMBC's Atmospheric Physics department would like to study the changes in attribute values such as the change in the value of humidity over time or the change in wind speeds between subsequent timesteps. For example, the three left images in Figure 7.21 show three timesteps of the humidity attribute in a hurricane. The change in humidity values

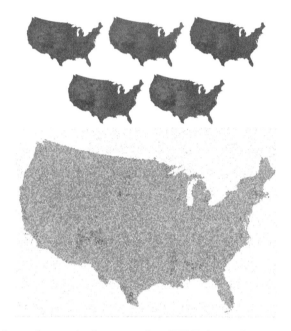

FIG. 7.19. This image shows a visual representation of US Voting results per county. The top two rows show county results from 1988, 1992 and 1996, 2000 and 2004. The pointillistic visualization clearly shows interesting patterns such as the fact that a section in Texas consistently voted Democrat over the entire 16 year interval.

is not apparent by looking at these snapshots. There is clearly a change in humidity values, but specific regions are not clear. Using standard visualization techniques, one can find the difference between the interval and use a color mapping to convey the change, as in the rightmost image in Figure 7.21. This visualization only manages to capture the overall change: it does not capture the values of the intermittent timesteps and therefore, does not give a complete representation of the variation in attribute values that the hurricane has undergone across the interval.

We applied our technique to various attributes of a hurricane, such as humidity and wind speeds. Figure 7.22 depicts three various painterly visualizations that were created using our system. The leftmost image depicts a single timestep depicted in a painterly style. The mid-

FIG. 7.20. This image shows a visual representation of US Voting results per county. The top row shows snapshots of election results from 1960, 1964 and 1968. The pointillistic visualization clearly shows variation in the southeastern region.

dle image depicts three timesteps with decreasing brush size and saturation; the third image depicts five timesteps with decreasing brush sizes and saturation. The rightmost image shows the size/saturation legend for the visualizations.

The third image from the left shows the final painterly visualization with five timesteps of humidity varying over a 15-hour time interval. The sampling frequency for the intermediate brush strokes is every three hours. The trend to note in this case is that the older humidity values are much lower, as depicted by various shades of desaturated, small blue and green brush strokes on the right of the eye of the hurricane, which is moving leftward. The current timestep has higher humidity, as depicted by the purple, violet, and dark blue large brush strokes. Homogeneity in the values towards the lower right region of the hurricane can be seen in the form of smaller, desaturated brush strokes. When viewed from a distance, the overall visualization conveys information regarding the humidity distribution over the interval, and a zoomed-in view of the data can give the user details

132

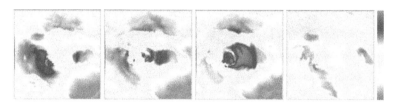

FIG. 7.21. Visualization of a 2D slice of the humidity attribute in a hurricane. The images from left to right show the humidity attribute at timestep 29, timestep 35, timestep 41 and the computed difference in humidity between timestep 29 and timestep 41.

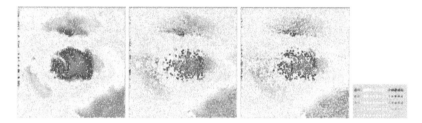

FIG. 7.22. Three different visualizations with different time intervals. The left image depicts humidity over a single timestep, which explains the constant brush size used in the image. The second image depicts humidity over three timesteps; smaller, desaturated brush strokes can now be seen interspersed with bigger brush strokes. The third image from the left depicts humidity over five timesteps and shows more variation in the form of older, desaturated brush strokes. The rightmost image shows a legend that depicts brush size and saturation values for viewers to correlate brush colors with values and timesteps.

regarding the humidity values of older timesteps.

The leftmost image in Figure 7.22 seems denser than the others to its right. This is due to the fact that the brush strokes are all of the same size and saturation, because only humidity values from a single timestep are being visualized. In the other images, we introduce information from older timesteps in the form of smaller, desaturated brush strokes that "compete" for the canvas. The effect seems more diffused as we look at the second and third image from the left in Figure 7.22 and the contribution of older timesteps becomes apparent in the painterly visualization.

Figure 7.23 visualizes changes in wind speed over a 20-hour interval, sampled every four hours for the painterly visualization. Here we investigate regions surrounding the eye of the hur-

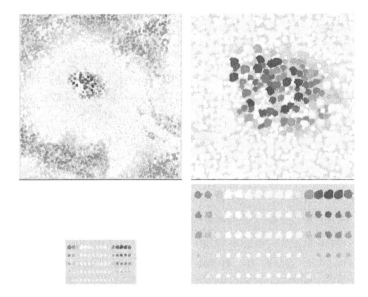

FIG. 7.23. Painterly visualization of wind speeds at different zooming levels. The left column shows the humidity attribute and a corresponding legend. The image depicts large wind speeds and change around the eye. On zooming in further, the right column depicts further diversity around the eye, the corresponding legend helps to identify values and variation over the interval. Zooming into the eye conveys to the user the fact that the eye contains the highest speed winds in the dataset,as depicted by the violet and dark blue brush strokes.

ricane and inspect them further by zooming into the region of interest. Notice that the legend is regenerated to match the size of the brush strokes being drawn on the canvas.

As expected, the change in wind speeds in and around the eye is much higher than farther away from the eye. In the lower left corner of the hurricane, various brush strokes colored with various shades of red depict smaller values and homogeneity in wind speed in those regions. On zooming into the hurricane visualization, as shown in the right image in Figure 7.23, one can see the high windspeed values depicted by violet brush strokes in and around the eye of the hurricane.

The trend to notice in this visualization, is that the wind speed values in older timesteps are higher in the lower right regions, while the upward leftward motion of the hurricane causes bright yellow brush strokes to be placed in the lower right region of the image. There are small, faded green strokes among large bright yellowish orange brush strokes in that region. The wind speed values in the eye of the hurricane are high even in older timesteps as can be seen in desaturated, smaller purple, violet brush strokes around the right side of the zoomed in image around the center.

7.3   Discussion

We believe that our techniques for encoding temporal attribute change produce an aesthetically pleasing visualization of the underlying data. The ability to show variations in time-varying data such monotonic increase and decrease, as an increase followed by a decrease or vice versa in a single visualization are clearly the strengths of the technique. The "visual blending" concept works well to allow viewers to view the visualization from a distance to get an overview. It helps to identify regions of change as well as trends in the data and facilitates easy investigation of those regions by zooming into those regions. The automatically updated legend greatly helps in the understanding of the generated pointillistic visualization.

As can be expected, the amount of information that can be usefully represented will decrease as the number of timesteps being considered increases, due to the limited amount of space on the canvas. We have found the number of timesteps that can be sampled and effectively visualized to be around five. We believe that there is an upper bound for the number of timesteps that can be visualized to effectively show change over a time interval.

One of the limitations of our algorithm is the range of brush sizes that can be effectively used. Using a brush size larger than 10 clutters the canvas. Using the smallest brush size of 2 is not useful, since we found that users cannot identify the desaturated brush strokes at that size. This can be seen in Figure 7.24, which shows the effect of specifying the minimum brush size as 2.

The smallest effective brush stroke size is of size 4 and the largest of size 8, thus allowing us to effectively visualize five timesteps. We could visualize more than five timesteps and increase our largest brush stroke to size 10, but as mentioned before, we run into visual clutter that may cause information overload.

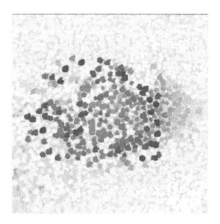

Fɪɢ. 7.24. The image depicts a closeup of humidity when the minimum brush size is 2. The smallest brush strokes can barely be seen due to its size and desaturation especially around the center and towards the bottom where there are faded blue brush strokes of size 2. For a meaningful visualization, the minimum brush size must be clearly visible and allow a viewer to correlate values with the legend. The brush sizes in the legend for minimum brush size = 4 are already small.

Chapter 8

EVALUATION OF POINTILLISM-BASED TECHNIQUES

This chapter describes the second of two user studies that we conducted to evaluate our art-inspired techniques. In this particular study, we evaluated the effectiveness of pointillism-based techniques in conveying change in value variant time-varying data. To evaluate the effectiveness we designed a user study that required our subjects to perform tasks similar to that of a domain expert. User accuracy, time required to complete each task and confidence in their answers were used to evaluate the pointillism-based techniques.

8.1 Hypothesis

Our hypothesis was that the new pointillism-based visualization techniques was more effective at visualizing trends in time-varying data than standard snapshots and animation-based techniques.

8.2 Independent variables

The independent variable in our case was the visualization technique. The visualization techniques were used to visualize a mix of synthetic and real-world datasets. The visualization technique was one of:

- Images - A panorama of snapshots for every timestep in the data. The number of snapshots used were five.

- Pointillism-based visualization depicting change over time. For this visualization, the number of timesteps considered to generate the visualization was five.

- Animation showing snapshots at each timestep, one after the other. The number of frames in the animation correspond to the number of timesteps, which was restricted to five.

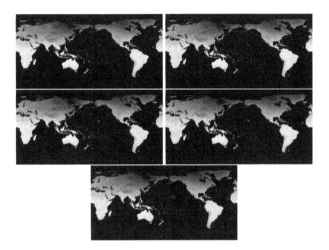

FIG. 8.1. Five snapshots showing the global temperature for each decade from 1950 to 2000. Greyscale intensity conveys the magnitude of temperature in regions.

Figure 8.1 shows five snapshots of temperature measured over fifty years. Each snapshot represents the temperature over a decade. Figure 8.2 shows a pointillism-based visualization of the same data.

8.3 Pilot experiments

Before I began the formal evaluation process, I ran a pilot experiment. The study was not timed or scored for user accuracy. The pilot study was conducted with four subjects whose answers were not considered in the final evaluation of the techniques.

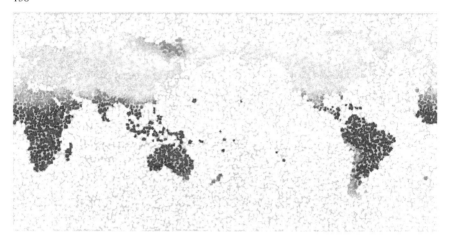

FIG. 8.2. A pointillism-inspired visualization of the global temperature data for the time interval of 50 years from 1950 to 2000.

As is with most pilot studies (2003), I found some problems with our study which were fixed before we began the formal evaluation. The problems that were identified were as follows:

- The questions based on identifying states required subjects to have general information regarding the states. Based on feedback from subjects, I added another map of the United States that would help subjects identify correct state names.

- Situations where the task might lead to more than one correct answers were fixed by changing the task or by replacing that task with a more unambiguous task.

- Subjects were inconsistently interpreting some tasks they were required to perform. Expanding task descriptions solved that problem.

- Incorrect options for the task being performed by the subjects. In a couple of cases the subjects could not complete the task, since there was no correct answer for the task they were performing.

- Typographical errors in task descriptions and questions were identified and fixed.

Additionally, for the pointillism-based evaluation, all the subjects were trained with the technique and sample example visualizations. Details of the training are provided in Appendix A.

8.4 Subjects

We tested our techniques with 24 subjects who had basic familiarity with using computers. We did not restrict ourselves to any age group or gender. We performed full factorial, within-subjects testing to evaluate our techniques. In order to balance the ordering effects, we randomly assigned subjects to trial orders.

Table 8.1 shows the ordering used for the subjects. We presented the subjects with snapshots, pointillism-based visualizations, and animations. The ordering used within each category is provided in Table 8.2. Table 8.4 shows the entire table for the study enumerating the order in which each user would see each visualization.

During the user study, we followed the following procedure:

1. Explain the user study to them and inform them of what it entails.

2. Obtain consent from the subjects for the user study.

3. Present them with training material and let them familiarize themselves with the task for some time.

4. Conduct the user study and collect the results after they were done

5. Present a usability questionnaire and get subjective feedback from the subjects. Questions for this questionnaire are specified in the next section.

140

Sub ID	Snapshot	Pointillism	Animation	Pointillism (Trends)
1	1	2	3	4
2	1	2	4	3
3	1	3	2	4
4	1	3	4	2
5	1	4	2	3
6	1	4	3	2
7	2	1	3	4
8	2	1	4	3
9	2	3	1	4
10	2	3	4	1
11	2	4	1	3
12	2	4	3	1
13	3	1	2	4
14	3	1	4	2
15	3	2	1	4
16	3	2	4	1
17	3	4	2	1
18	3	4	1	2
19	4	1	3	2
20	4	1	2	3
21	4	2	1	3
22	4	2	3	1
23	4	3	2	1
24	4	3	1	2

Table 8.1. Ordering for evaluation of our techniques in comparison to standard visualization techniques.

a	b	c	d
a	b	c	d
a	b	d	c
a	c	b	d
a	c	d	b
a	d	b	c
a	d	c	b
b	a	c	d
b	a	d	c
b	c	a	d
b	c	d	a
b	d	a	c
b	d	c	a
c	a	b	d
c	a	d	b
c	b	a	d
c	b	d	a
c	d	a	b
c	d	b	a
d	a	b	c
d	a	c	b
d	b	a	c
d	b	c	a
d	c	a	b
d	c	b	a

Table 8.2. In each of the four categories mentioned in table 8.1, the subject was shown four different images or animations. These examples were varied according to the order given here.

ID	Snapshot				Pointillism				Animation				Trends			
1	1-a	1-b	1-c	1-d	2-a	2-b	2-c	2-d	3-a	3-b	3-c	3-d	4-a	4-b	4-c	4-d
2	1-a	1-b	1-d	1-c	2-a	2-b	2-d	2-c	4-a	4-b	4-d	4-c	3-a	3-b	3-d	3-c
3	1-a	1-c	1-b	1-d	3-a	3-c	3-b	3-d	2-a	2-c	2-b	2-d	4-a	4-c	4-b	4-d
4	1-a	1-c	1-d	1-b	3-a	3-c	3-d	3-b	4-a	4-c	4-d	4-b	2-a	2-c	2-d	2-b
5	1-a	1-d	1-b	1-c	4-a	4-d	4-b	4-c	2-a	2-d	2-b	2-c	3-a	3-d	3-b	3-c
6	1-a	1-d	1-c	1-b	4-a	4-d	4-c	4-b	3-a	3-d	3-c	3-b	2-a	2-d	2-c	2-b
7	2-b	2-a	2-c	2-d	1-b	1-a	1-c	1-d	3-b	3-a	3-c	3-d	4-b	4-a	4-c	4-d
8	2-b	2-a	2-d	2-c	1-b	1-a	1-d	1-c	4-b	4-a	4-d	4-c	3-b	3-a	3-d	3-c
9	2-b	2-c	2-a	2-d	3-b	3-c	3-a	3-d	1-b	1-c	1-a	1-d	4-b	4-c	4-a	4-d
10	2-b	2-c	2-d	2-a	3-b	3-c	3-d	3-a	4-b	4-c	4-d	4-a	1-b	1-c	1-d	1-a
11	2-b	2-d	2-a	2-c	4-b	4-d	4-a	4-c	1-b	1-d	1-a	1-c	3-b	3-d	3-a	3-c
12	2-b	2-d	2-c	2-a	4-b	4-d	4-c	4-a	3-b	3-d	3-c	3-a	1-b	1-d	1-c	1-a
13	3-c	3-a	3-b	3-d	1-c	1-a	1-b	1-d	2-c	2-a	2-b	2-d	4-c	4-a	4-b	4-d
14	3-c	3-a	3-d	3-b	1-c	1-a	1-d	1-b	4-c	4-a	4-d	4-b	2-c	2-a	2-d	2-b
15	3-c	3-b	3-a	3-d	2-c	2-b	2-a	2-d	1-c	1-b	1-a	1-d	4-c	4-b	4-a	4-d
16	3-c	3-b	3-d	3-a	2-c	2-b	2-d	2-a	4-c	4-b	4-d	4-a	1-c	1-b	1-d	1-a
17	3-c	3-d	3-a	3-b	4-c	4-d	4-a	4-b	1-c	1-d	1-a	1-b	2-c	2-d	2-a	2-b
18	3-c	3-d	3-b	3-a	4-c	4-d	4-b	4-a	2-c	2-d	2-b	2-a	1-c	1-d	1-b	1-a
19	4-d	4-a	4-b	4-c	1-d	1-a	1-b	1-c	2-d	2-a	2-b	2-c	3-d	3-a	3-b	3-c
20	4-d	4-a	4-c	4-b	1-d	1-a	1-c	1-b	3-d	3-a	3-c	3-b	2-d	2-a	2-c	2-b
21	4-d	4-b	4-a	4-c	2-d	2-b	2-a	2-c	1-d	1-b	1-a	1-c	3-d	3-b	3-a	3-c
22	4-d	4-b	4-c	4-a	2-d	2-b	2-c	2-a	3-d	3-b	3-c	3-a	1-d	1-b	1-c	1-a
23	4-d	4-c	4-a	4-b	3-d	3-c	3-a	3-b	1-d	1-c	1-a	1-b	2-d	2-c	2-a	2-b
24	4-d	4-c	4-b	4-a	3-d	3-c	3-b	3-a	2-d	2-c	2-b	2-a	1-d	1-c	1-b	1-a

Table 8.3. Complete table for user evaluation of pointillism-based techniques

8.4.1 Datasets

We evaluated our techniques with a combination of synthetic and real-world data. Synthetic data were generated to contain all the possible trends that can be conveyed using our techniques. We generated synthetic time-varying datasets in which the values vary over time. The datasets were generated to evaluate whether our pointillism-based techniques were able to show trends over time.

Real-world data was used to test the effectiveness of the techniques and compare it with standard visualization techniques. We evaluate our techniques on real-world datasets that we have obtained from various sources:

- Hurricane data obtained from NASA courtesy of Dr. Lynn Sparling and Dr. Scott Braun.

- Rainfall data for the world for the last 100 years, obtained from NOAA's website (NASA 2000).

- US infant mortality per state for the last 10 years from the US Census bureau (CDC 2002).

- Voting results per county from 1960-2004 for the entire nation (PurpleAmerica 2004).

8.5 Tasks

The tasks were based on the kind of tasks that an application domain scientist would perform on a regular basis to correlate timestep data. The tasks require a user to identify variability and various temporal trends in a time-varying dataset by examining a visualization.

The users were asked to perform the task of identifying a trend in a highlighted region in the dataset. The user was asked to correctly identify whether the trend in the data was increasing, decreasing, increasing and decreasing, decreasing and increasing or constant over time.

The user was shown the following:

- Snapshots of subsequent timesteps (Baseline - standard visualization technique).

- An animation showing the snapshots sequentially (Baseline).

- Our pointillism-based visualization of the time-varying data.

FIG. 8.3. This set of diagrams represent the kinds of temporal trends that we tested our techniques. The images from left to right, show an increasing, a decreasing, a constant, a decrease followed by an increase, an increase followed by a decrease. The subjects were asked to identify a trend using such diagrams.

The subject were shown small diagrams indicating the possible trend in the specified region and the subject then picked a trend based on their observation. The time required to complete the task, accuracy, and confidence of the subjects in their answers were used to evaluate the visualization techniques.

8.5.1   Evaluation questions

The questions in the user evaluation were based on identifying trends. We asked questions such as:

- Based on the presented visualization, identify the trend in the data. Does it seem to be monotonically increasing, monotonically decreasing, an increase followed by a subsequent decrease, a decrease followed by a subsequent increase, or no change?

- In the shown visualization, identify regions of variability and high change.

8.5.2   Subjective evaluation

The subject was requested to fill out a questionnaire to evaluate where the usability. Their answers were obtained using a scale from 1 (easy/agree) to 9 (hard/disagree). Questions asked in the questionnaire were as follows:

1. Were the questions asked for each evaluation straightforward?

2. Was it easy to perform simple tasks such as identifying variability/change using standard visualization techniques (snapshots and animations)?

3. Was it easy to perform tasks such as identifying trends using standard visualization techniques (snapshots and animations) ?

4. Could you perform simple tasks such as identifying a single trend using pointillism-based techniques?

5. Was it easy to perform tasks such as identifying trends using pointillism-based techniques?

6. Did you prefer the pointillism-based techniques or the standard snapshots/animation techniques?

8.6 Dependent Variables

I use the user performance time, user accuracy and confidence to evaluate the techniques. *User performance time* is the time required to complete each task. *User accuracy* indicates whether the use of our techniques helped the users to answer the questions correctly. I also measured the confidence of the subjects in their answers. In order to compute *user subjective satisfaction*, I asked subjects to rate their experience on a scale of 1-9 for a variety of questions.

8.7 Conducting the user study using a web browser

The user study was conducted using a web browser. Since I had to show videos to the subjects, we used Riva Free FLV encoder (RIVA 2006) to encode AVI files into flash files that can be shown to the viewer in the browser setting. Figure 8.4 is a screenshot of the first screen shown to the subjects. It gives an overview of the user study to the subjects. After answering any other questions that the subject had, we obtain the consent of the subject to continue the user study.

Figure 8.5 shows a screenshot of a screen that the subjects saw. The user was asked to identify a trend or variability in the data using one of the visualization techniques. The user then observes it and provides an answer. The user also indicates a confidence level in their answer.

146

FIG. 8.4. A screenshot of the welcome screen that was shown to the subjects. The subjects were explained the user study and any other questions they may have were answered before obtaining their consent.

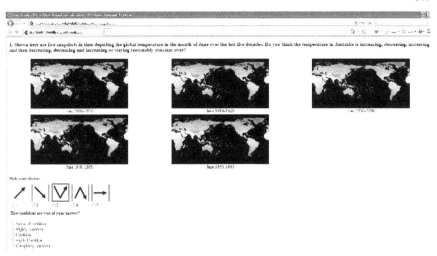

FIG. 8.5. A screenshot of a sample screen that is shown to the subjects. The subjects were shown images and they were asked to indicate the overall change that they can visually identify in the data. They were also requested to give their confidence level in their answer.

8.8 Results

The ability of the subjects to visualize variability in a region over time and their ability to identify specific temporal trends was evaluated. High *variability* corresponds to large changes in a region over time; low variability corresponds to static values that do not seem to change over time. Our techniques can also convey various temporal *trends*. A temporal trend could be increasing values of rainfall in a region over a given time interval.

To compare the three different visualization types we use the statistical test *Analysis of Variance (ANOVA)*. This test allows us to compare the accuracy, timings and confidence obtained from the three groups (snapshots, animations, and pointillist techniques). The test begins with a null hypothesis that the use of pointillism-based techniques provides no speedup in completing tasks, no improvement in accuracy, and that the users feel equally confident in their answers for all techniques. The statistical measure of significance p evaluates the probability of the result agreeing with the null hypothesis. For values of $p < 0.05$, the null hypothesis is rejected, implying that the

Type of visualization	Mean	Standard Deviation
Snapshots	75.556	15.967
Animations	81.481	12.346
Pointillism	88.889	25.514

Table 8.4. This table shows the mean and standard deviation of the accuracy of the users. The pointillism-based technique helped subjects complete the task more accurately.

use of pointillism-based techniques makes a difference.

8.8.1   Variability results and analysis

The ability of the subjects to notice variability in a region over time was evaluated and analyzed. Table 8.4 shows the mean and standard deviation for the average user accuracy for all users over all questions, using each visualization category. The mean accuracy for the pointillism-based techniques was higher than that of animations as well as snapshots. Figure 8.6 shows a graphical representation of the results. It shows that subjects were more accurate when using the pointillism-based techniques than with snapshots or animations. The snapshots and the animations require the user to correlate changes and observe a large region with continuously changing values over time. Our pointillism-based technique produces a visual representation that obtains contributions from all the intermediate time intervals. It highlights regions of low as well as high variability, allowing the subjects to accurately complete the task.

Table 8.5 shows the result of analyzing the user accuracy using ANOVA. The computed probability p of this result assuming the null hypothesis was less than 0.029. This implies that the result was *statistically significant* and that the null hypothesis was rejected. The pointillism-based techniques increase the ability of a user to accurately visualize trends in time-varying data.

We also measured and analyzed the time required for subjects to complete each task. Table 8.6 shows the mean and standard deviation values for the timings results for the subjects. Figure 8.7 shows a graphical representation of the timing results. In this case, the subjects required much less time to complete the task than with snapshots or animations. In the case of snapshots, this may be due to the fact that the subjects have to correlate values and interpret changes using all the snapshots and the color scale. The animations technique seems to allow users to identify variability

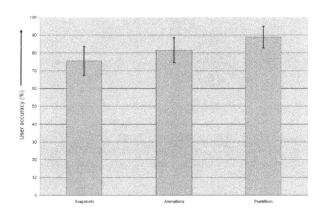

FIG. 8.6. This graph shows that the overall accuracy of the subjects was higher for pointillism-based visualization than for snapshots or animations.

Source of variation	Sum of squares	Degrees of freedom	Mean squares	F
between	2410.	2	1205.0	3.719
error	2.5274E+04	78	324.0	
total	2.7684E+04	80		

Table 8.5. This table shows the result of performing the ANOVA test on the accuracy per user. The probability of this result, assuming the null hypothesis, was less than 0.029. This implies that the result was statistically significant: the null hypothesis(that the pointillism-based techniques do not increase the accuracy in completing a task) was rejected.

Type of visualization	Mean	95% confidence interval	Standard Deviation
Snapshots	102.67	79.36-126.0	9.234
Animations	91.533	68.23-114.8	10.044
Pointillism	65.585	42.28-88.89	8.722

Table 8.6. This table shows the mean, 95% confidence intervals around the mean and standard deviation for the timings results.

Source of variation	Sum of squares	Degrees of freedom	Mean squares	F
between	3.2667E+04	2	1.0889E+04	3.316
error	3.4147E+05	78	3283.0	
total	3.7414E+05	80		

Table 8.7. This table shows the result of performing the ANOVA test on the time required by the users to complete a task using all the techniques. The probability of this results, assuming the null hypothesis, was less than 0.023. This implies that the result was statistically significant and the null hypothesis that the pointillism-based techniques do not affect the speed of the subjects in completing a task, was rejected.

faster, but not as fast as pointillism-based techniques. Therefore, not only did the subject require less time to finish the task using pointillism-based techniques, but also were more accurate, as can be seen in Table 8.4 and in Figure 8.6.

On analyzing the timings over all the users using ANOVA, as shown in Table 8.7, we found that probability p of the null hypothesis was 0.023. Since the probability was less than 0.05, the result was *statistically significant*. This implies that the pointillism-based techniques aid the ability of the subjects to complete the task quickly, compared to snapshots or animations.

Along with accuracy and time required per user per task, we requested the subjects to state their confidence in their answers for each question. Table 8.8 shows the mean and standard deviation in user confidence for the three visualization techniques. Figure 8.8 shows a graph depicting the average confidence. Users were more confident about their answers for pointillism-based techniques than for snapshots or animation-based techniques. The pointillism technique was ideal for visualizing variability in data and a single representative image can convey low as well as high variability. Our pointillism-based techniques draw the viewer's attention to regions of high variability and seem to instill more confidence in the subjects than do snapshots or animations.

Performing the ANOVA test on the confidence over all answers, as shown in Table 8.9, the

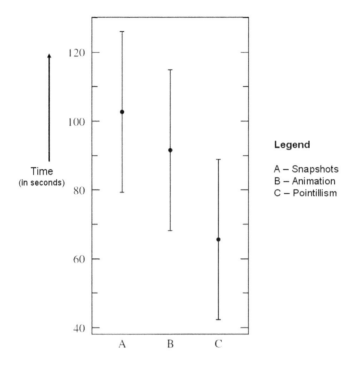

FIG. 8.7. This graph shows the mean time required by subjects to complete the tasks using snapshots (A), animations (B), and pointillism-based techniques (C). The errors bars show the 95% confidence interval around the mean. The pointillism-based techniques takes almost 35 seconds less than snapshots and 25 seconds less than animation, on average.

Type of visualization	Mean	95% confidence interval	Standard Deviation
Snapshots	3.5407	3.228-3.854	0.135
Animations	3.4889	3.176-3.802	0.122
Pointillism	3.711	3.398-4.024	0.129

Table 8.8. This table shows the mean, 95% confidence intervals around the mean as well as standard deviation for the confidence results.

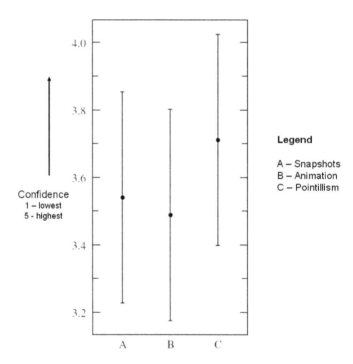

FIG. 8.8. This graph shows the confidence the subjects had in their answers. The users seem more confident of their answers when using pointillism. They also seem to be more confident of their answers with snapshots than with animations.

Source of variation	Sum of squares	Degrees of freedom	Mean squares	F
between	8.080	2	2.693	3.663
error	76.47	78	0.7352	
total	84.55.0	80		

Table 8.9. This table shows the result of performing the ANOVA test on the confidence per user per answer. The probability of this results, assuming the null hypothesis, was less than 0.015. This implies that the result is statistically significant and the null hypothesis that the pointillism-based techniques do not increase the accuracy in completing a task, was rejected.

Type of visualization	Mean	Standard Deviation
Snapshots	66.667	34.174
Animations	74.275	22.128
Pointillism	77.875	25.514

Table 8.10. This table shows the mean and standard deviation of the accuracy of the users when visualizing temporal trends. Subjects seem to be more accurate when using the pointillism-based technique than with snapshots and only marginally better when using animations.

computed probability p of the null hypothesis being true was 0.015. Since $p < 0.05$, the null hypothesis was rejected and a *statistically significant* result was obtained. This implies that the increased user confidence in pointillism-based techniques was highly unlikely to have occurred by chance.

8.8.2 Trends results and analysis

The ability of the subjects to visually correlate and identify trends in data was evaluated. The accuracy, time required per task per technique and the user confidence was measured and analyzed.

We analyzed whether the subjects were able to accurately identifying trends in the time-varying data using our pointillism- based techniques compared to snapshots and animations. Table 8.10 shows the mean and standard deviation in the accuracy of the subjects. Figure 8.9 shows a graph of the result. The accuracy using pointillism-based techniques was higher than snapshots, but only marginally higher than animations. Users seem to be able to visualize trends in data well using animations too.

Performing the ANOVA test on the user accuracy over all the users over all the tasks, we get the results as shown in Table 8.11. The computed probability p of the null hypothesis being true

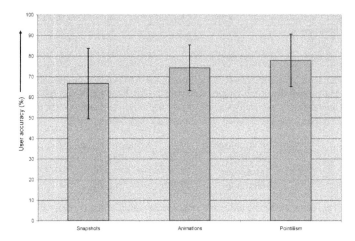

FIG. 8.9. This graph shows the accuracy of the subjects when completing tasks requiring them to find temporal trends in the data. Subjects seem to be more accurate when using pointillism based techniques than with snapshots, but only slightly more accurate than with animations.

Source of variation	Sum of squares	Degrees of freedom	Mean squares	F
between	3350.2	2	1402.6	2.716
error	3.2154E+04	78	472.2	
total	2.9823E+04	80		

Table 8.11. This table shows the result of performing the ANOVA test on the user accuracy per user per answer. The probability of this results, assuming the null hypothesis, was less than 0.024. This implies that the result was statistically significant and the null hypothesis that the pointillism-based techniques do not increase the accuracy in completing a task, was rejected.

was 0.024. Since $p < 0.05$, the null hypothesis was rejected and a *statistically significant* result was obtained. This implies that the pointillism-based technique helps increase the user accuracy.

Additionally, the time required per user to complete each task per visualization technique was measured and analyzed. Table 8.12 shows the mean and standard deviation of the time required for each visualization technique. Figure 8.10 shows a graph showing the mean and the 95% confidence interval around it. The time required for subjects to find trends was significantly lower than that using snapshots or animations. The ability to visualize values from various timesteps and correlating them in a single image seems to provide subjects with sufficient information to identify

Type of visualization	Mean	95% confidence interval	Standard Deviation
Snapshots	101.00	74.06-127.9	10.541
Animations	92.719	65.78-119.7	9.42
Pointillism	61.44	34.20-88.08	11.089

Table 8.12. This table shows the mean, 95% confidence intervals around the mean as well as standard deviation for the timings results.

Source of variation	Sum of squares	Degrees of freedom	Mean squares	F
between	2.3886E+04	2	1.1943E+04	2.416
error	3.8563E+05	78	4944.0	
total	4.0952E+05	80		

Table 8.13. This table shows the result of performing the ANOVA test on the timing per user per answer. The probability of this results, assuming the null hypothesis, was less than 0.046. This implies that the result was statistically significant and the null hypothesis that the pointillism-based techniques do not increase the speed in completing a task, was rejected.

trends faster.

The time required per user over all the answers was analyzed using the ANOVA test as shown in Table 8.13. The computed probability p of the null hypothesis being true was 0.046. Since $p < 0.05$, the null hypothesis was rejected and a *statistically significant* result was obtained. This implies that the pointillism-based technique helps subjects complete a task faster.

The user confidence per task was obtained and analyzed. Table 8.14 shows the mean and standard deviation for the confidence that the users had in their answers for identifying trends. Figure 8.11 shows a graph of the result. The subjects were more confident in their answers for animations and snapshots than with the pointillism-based techniques. This can be attributed to the fact that the pointillism-based technique was unfamiliar to the subjects and so even though their answers were correct, as was verified, the subjects were just less confident than with animations and snapshots.

The ANOVA test results for the confidence answers are shown in Table 8.15. The computed probability p of the null hypothesis being true was 0.078. Since $p > 0.05$, the null hypothesis cannot be rejected and so we cannot deduce that a statistically significant result was obtained. The result indicates that the high confidence in snapshots and animations as compared to the pointillism-based technique could have occurred by chance and no technique clearly instills more confidence

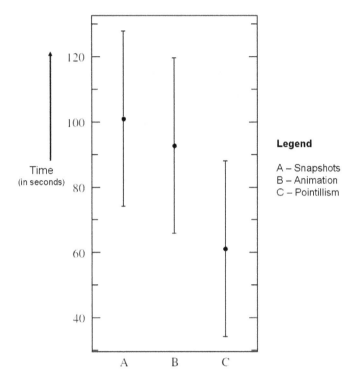

FIG. 8.10. This graph shows the confidence the subjects had in their answers. As per the graph, the users seem more confident of their answers when using pointillism. They seem to be more confident of their answers in the case of snapshots as compared to animations.

Type of visualization	Mean	95% confidence interval	Standard Deviation
Snapshots	3.4370	3.119-3.755	0.127
Animations	3.711	3.393-4.030	0.098
Pointillism	3.1926	2.874-3.511	0.113

Table 8.14. This table shows the mean, 95% confidence intervals around the mean as well as standard deviation for the confidence results.

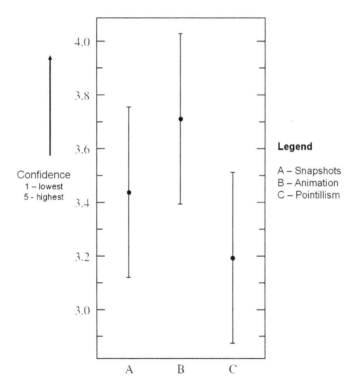

FIG. 8.11. This graph shows the confidence the subjects had in their answers. As per the graph, the users seem more confident of their answers when using animations and snapshots than when using pointillism. This could be attributed to the fact that the users were unfamiliar with the pointillism-based technique as compared to animations and snapshots.

Source of variation	Sum of squares	Degrees of freedom	Mean squares	F
between	3.634	2	1.817	2.631
error	53.87	78	0.6906	
total	57.50	80		

Table 8.15. This table shows the result of performing the ANOVA test on the confidence per user per answer. The probability of this results, assuming the null hypothesis, was less than 0.078. This implies that the result was not statistically significant and the null hypothesis cannot be rejected.

as compared to the other techniques.

8.9   Discussion

Analyzing the results from the user study provided some insights into the effectiveness of my pointillism-based technique. The aim was to evaluate whether the technique was useful in conveying variability in time-varying data and whether it was able to convey trends in the data.

Particularly, in the case of visualizing variability, our pointillism-based technique was more effective than snapshots and animations, because it samples all the timesteps in the time interval to create a visualization. This ensured that the variability was captured effectively. My hypothesis was validated by the fact that the subjects were able to perform variability-based tasks more accurately and quickly than with snapshots or animations. In the case of snapshots, it can be explained by the fact that the subjects have to observe each snapshot, correlate the data values in that snapshot with the color legend, and then infer variability or identify a pattern from it. This takes more time than looking at a single, comprehensive image. Comparatively, in the case of animations, users required less time than snapshots, since our visual system is adept at identifying variability. The fact that the subjects had to look at the complete animation before they could answer questions might have affected the time required by the subjects to complete the task. This is naturally a disadvantage for animations when visualizing time-varying data over a larger time interval. Considering that our pointillism-based technique can effectively visualize only five timesteps as of now, this problem of viewing time-varying data over a larger interval still needs to be addressed.

When the users were asked to identify trends in the data using our pointillism-based techniques, they performed marginally better than animations and much better than snapshots in terms of accuracy and speed, but were not as confident in their answers. This may be due to the fact that these techniques are new and the users were not sure whether they were reading the visualizations correctly. This was supported by their answers in the subjective evaluation, where a majority of them said that it was less straightforward to identify trends using our pointillism-based techniques. Considering that they thought that it was harder to identify trends using snapshots and animations implies that the task was hard to begin with. The fact that their accuracy was reasonably high implies that they were just not confident of their correct answers.

In the case of identifying specific trends, the users' performance did not vary by the trend they were asked to identify. Users performed equally well at identifying the simpler trends, such as monotonic increase or monotonic decrease, and harder trends, such as an increase followed by a decrease or decrease followed by an increase. From the pilot study I had found that identifying more complex trends seemed to be a problem, but over a larger population of the user study, it did not affect the subjects: they performed as well when identifying simple and complex trends.

In order to identify, whether the subjects got better at visualizing trends using pointillism, I analyzed the results of their accuracy based on the user evaluation data. Out of the entire subject population, 9 subjects displayed improvements in identifying trends while 3 displayed a degradation in their accuracy results. It implies that some subjects clearly improved as they answered more questions using the pointillism-based visualization.

The subjective evaluation results implied that the users preferred to use the pointillism-based techniques over snapshots and animations to visualize the variability. They found it hard to identify trends in data using snapshots as well as animations. Four users, in particular, performed consistently badly on identifying trends with all snapshots, animations and pointillism-based techniques. In the final question, where they were asked whether they preferred the pointillism-based technique over snapshots and animations, a score of 3.1 was received on a scale of 1 (agree) to 9 (disagree). The accuracy, timing, and confidence, and subjective evaluation indicates that the users seem to benefit from the use of pointillism-based technique in completing tasks for identifying variability and trends in time-varying data.

Chapter 9

CONCLUSIONS AND FUTURE WORK

In this work, I have presented novel art-inspired techniques to visualize time-varying data. Using my techniques, application domain users can visualize time-varying data more effectively.

This work is valuable to domain experts who would prefer to watch temporal summaries of time-varying data. Alternatively, they may have to watch hours of animations/surveillance video which may have no interesting activities. A temporal summary can provide an overview of changes over time and allow the expert to visualize temporal data effectively.

9.1 Illustrative cues for visual tracking

Time-varying data visualization using snapshots and animations of snapshots are limited at conveying change in position and change in attribute values. For time-varying data, where the features are three-dimensional and moving over time, our illustration-inspired techniques are useful in helping a user to visually track a feature over time. I have identified and applied our techniques for visualizing time-varying data using speedlines, flow ribbons, opacity-modulation, and strobe silhouettes to convey temporal change in feature positions over time.

Our user study helped us evaluate the effectiveness of our techniques in terms of the accuracy of the users conducting simple tasks, the amount of time required by the users to complete the task, and confidence in their answers. We found that our techniques helped subjects complete the task faster, more accurately, and with more confidence in their answers. Users preferred the use of speedlines over opacity-based techniques, apparently because in some cases they misinterpreted

160

opacity-based techniques as data from the same timestep.

In addition to identifying new techniques, we identified a novel intent-based simplification technique for curves and paths. The ability to generate simplified representations based on the need of the user is novel and can be applied in various domains such as hurricane paths, driving directions, maps, and illustrative visualization of paths in three-dimensional feature visualization.

9.2   Hurricane investigation through illustrative techniques

The application of illustration-inspired techniques to visualize the time-varying data from hurricanes Bonnie, Isabel, and Katrina led to interesting challenges and solutions. Our techniques have facilitated the study of hurricanes and allowed visualization of internal hurricane structures. Our main contribution was to demonstrate the use of illustration-based techniques, in a domain where 2D traditional visualization techniques have been used, to help understand the physics behind the phenomenon and assist the investigation of conditions leading to the intensification or dissipation of a hurricane. The silhouette computation method helped in the identification of vertical wind shear, which provides answers to crucial questions concerning the weakening and eventual dissipation of a hurricane. We provided domain experts with novel visualizations that provided temporal information regarding previous timesteps.

An expert evaluation was conducted to identify the effectiveness of our techniques. The experts mostly preferred the illustration-inspired techniques over standard visualization techniques. Visualizing energy flow through the entire hurricane system and its dependencies on assumptions about the coupling of the storm to the ocean can be further investigated to know more about the evolution of a hurricane.

9.3   Visualizing change in attribute values using pointillism

For visualizing changes in value variant time-varying data, we introduced novel pointillism-inspired techniques. Such changes in value are hard to visualize using standard methods and generally end up giving only an idea of overall change. Our technique uses a unique method of sampling every timestep to create a painterly visualization. We take advantage of the color mixing theory initially proposed by Rood (1879) and later used by Seurat in his paintings (Kemp 1990).

A user study was conducted to evaluate the ability of our pointillism-based techniques to convey change (variability) over time as well as visualize trends in the data. The user study indicated that users were able to visualize variability faster than snapshots and animations and with increased accuracy. The users were able to identify specific trends in the data using our pointillism-based technique faster than snapshots and animations. The accuracy was marginally better than animations, but the users were less confident of their correct answers than in the case of animations. This may have been due to the fact that the pointillism-based techniques were new compared to the animation technique, which the users were more familiar with.

9.4 Future work

There are several directions for future work. In the case of illustration-inspired techniques, identifying novel illustration-based techniques that can provide more information to the viewer could prove greatly useful and could add to the four techniques that have been introduced here.

The use of illustration-inspired techniques when considered with occlusions between objects and illustrative cues is worth exploring further. The complex interactions between multiple illustrative cues like speedlines and flow ribbons occluding each other. The effect on understanding the spatial relationships for situations when another feature is in front of a feature being visualized using illustration-inspired techniques, such as speedlines, needs to be further investigated. For example, in Figure 9.1 we can see four features of a synthetic dataset being augmented by speedlines. The direction of motion of the features is clear from the augmented visualization. The green feature towards the bottom right is spatial occluding the speedlines for the top green feature since it went over the path that the top green feature traversed previously. Such spatial relationships can be conveyed using visualizations such as these, but more complex, spatial relationships need to be further explored.

Intuitively, a large number of speedlines, strobe silhouettes, flow ribbons in a single image might make it hard to understand for a time-varying dataset with many features moving over time.

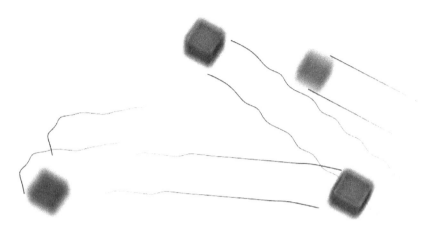

FIG. 9.1. This visualization shows four features in a synthetic dataset augmented with speedlines. The green feature in bottom right occludes the speedlines indicating the motion for the top green feature. This implies that the bottom green feature is traversing over the path that the top green feature took at an older time instant.

Further investigation needs to be performed into the number of cues that can be added in the case of multi-feature tracking. The boundary between visual clutter and effective visualization needs to be found for multi-feature tracking.

In the field of hurricane visualization, many other techniques could be identified by close collaboration with domain experts. In many cases, the solutions to their problems may not require a novel visualization technique, but it can make a significant impact in what the domain expert can see through the visualization. Dynamical systems such as these can be visualized using illustration-inspired techniques. The illustration-inspired techniques provide information regarding previous timesteps and the positional change in the dynamical system.

The pointillism-based technique though should not be used for positionally varying dynamical systems. The pointillism-based technique could be used to visualize spatial correlations in multi-attribute datasets. For example, correlations between vertical velocity and vorticity in a dynamical system could be effectively visualized using pointillism-based technique. The correlation could be visualized using higher-density of brush strokes in regions of high correlation while sparsely placed brush strokes could imply low correlation. Another interesting application of the pointillism technique could be to convey change in *land use* over time.

In the case of pointillism-based visualizations, users expressed that it was hard to distinguish brush strokes and correlate them with those in the palette. A perceptually constant color scale such as the LAB or LUV color space could be used instead of the HSV color space currently being used. Perceptually constant color scales may help users identify and correlate various brush colors more correctly. Using different shapes for each timestep was another user suggestion that could help the viewer to disambiguate closely placed brush strokes that look similar. From a distance, the brush strokes will blend as long as they are placed close enough.

TRAINING FOR EVALUATION OF POINTILLISM-BASED TECHNIQUES

A.1 Explanation of the visualization technique with simple examples

In this user study, we evaluate a new technique to visualize time-varying data. Let us look at a few examples to familiarize you with the technique. In the toy example shown in Figure A.1, the snapshots of each timestep are represented by five squares. The squares are colored according to the scale below. Looking at this set of five images, one can identify a pattern over time. To interpret this set of images, you have to use the color scale provided below.

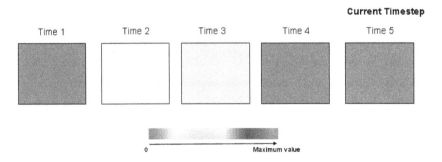

FIG. A.1. This figure shows five snapshots of a synthetic dataset. The color scale shown below helps visual correlation with values in each timestep.

The first snapshot on the left represented by "Time 1" is colored in Red, which corresponds

166

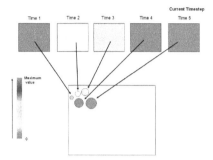

FIG. A.2. A schematic depicting how the brush size and brush color is chosen based on the *age* of the timestep.

to the value 0. The rightmost square for "Time 5" is colored violet which represents the maximum value. The squares in between have intermediate colors representing intermediate values. Therefore an increasing trend over these five timesteps can be clearly interpreted by this simple set of example images.

For a simple example, looking at a series of snapshots like in Figure A.1 is possible. But in real-world data, more complicated trends in values are observed. In some cases, the changes in values between timestep "t" and "t+1" may not be visually detected by an observer.

To allow an observer to visualize trends in time-varying data, we have developed a pointillism-based technique which is inspired by the painting paradigm. In our technique, we place brush strokes on the "canvas" to convey values from the timesteps. The color of the brush stroke is obtained by the value in that particular timestep. In order to differentiate, the brush stroke representing timestep "t" and timestep "t-1", we use brush size as well as brush brightness. Figure A.2 will help explain this better. In this image, the canvas at the bottom is filled with brush strokes by looking at the values in the five timesteps. The current timestep (violet square) is represented with the biggest brush stroke and maximum brightness. The older (t-1) timestep is represented with a smaller brush stroke and slightly reduced brightness and the oldest timestep in the series (Time 1) is shown with the smallest brush stroke and a very faded red brush stroke.

Using this technique, we can fill the entire canvas by randomly selecting one of the timesteps.

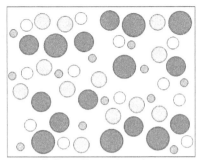

Pattern of increasing values

FIG. A.3. This image shows a visualization using our pointillism-based technique.

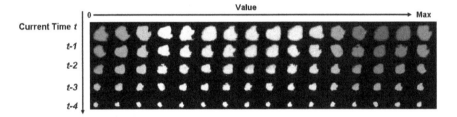

FIG. A.4. Color scale to be used for interpreting the pointillism-based visualization

Based on how old it is, the point-sized brush stroke with a certain color is placed on the canvas. Figure A.3 shows the canvas filled with brush strokes in this manner. Our technique generates the visualization shown in Figure A.3 for the simple example we saw in Figure A.1. The brush strokes used in the example are larger than normal, to convey the concept. We developed a new color scale to interpret the visualization. The color scale is shown in Figure A.4. Using the color scale, the brush strokes from the visualization can be mapped to a value and the brush size can convey its "age". The new color scale lets us correlate the values and infer the increasing trend in the data.

FIG. A.5. Five snapshots of synthetic data changing over time with the color scale used for the visualization.

A.2   Trend - Decreasing over time

In the simple example shown in Figure A.5, using the color scale shown on its right, we can infer that there is a decreasing trend .

Using our pointillism-based technique, we can convey this decreasing trend in a single image as shown in Figure A.6. The biggest brush stroke is yellow in color which implies it belongs to the current timestep. Older brush strokes in decreasing size and brightness can be seen. In decreasing size order, one can see a bright green stroke, a light bluish/green stroke, a sky blue stroke and the smallest violet brush stroke. Correlating it with the color scale helps you infer the trend. The 5 distinct brush strokes can be correlated with the brush scale shown below. The correlation helps us infer the decreasing trend.

A.2.1   Visualization generation process

Figure A.7 shows how more strokes get added to generate the pointillism-based visualization. The leftmost image shows the values from the current timestep with the largest, brightest brush stroke. As we go rightward, we can see increasingly smaller brush strokes being added to the visualization. As the values in the past timesteps were higher, the brush strokes picked are accordingly chosen. The color scale shown helps in the process of correlation.

169

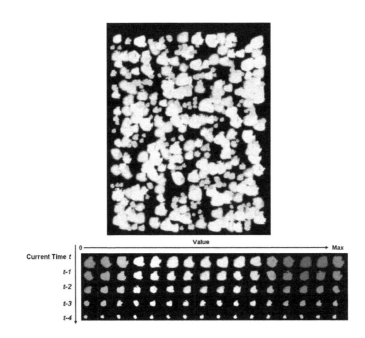

FIG. A.6. Five timesteps being visualizing using pointillism-based techniques in a single visualization. The color scale shown below helps the user identify trends in the data.

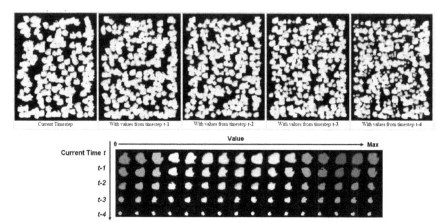

FIG. A.7. Five snapshots of synthetic data changing over time with the color scale used for the visualization.

A.3   Trend - Decrease followed by an increase

Figure A.8 shows an example of a trend where an decreasing trend is immediately followed by an increasing trend. The biggest brush stroke is blue in color, followed by a smaller bright green brush stroke, a subsequently lighter shade of green. At this point, the color scale on the right helps us infer that the values have been dropping. But then the other two brush strokes that can be seen are light greenish blue and violet in color. This implies that at older timesteps the value was high. Therefore a drop followed by a subsequent rise can be see in this image.

A.4   Trend - Constant value over time

Figure A.9 shows that the value is constant over the time interval. This can be seen by noticing different shades of green brush strokes. Starting from the bright big green brush strokes all the way to the smallest faded green brush strokes, they are all of the same color. Therefore, using the color scale shown below, we can infer the fact that the values are constant over time.

171

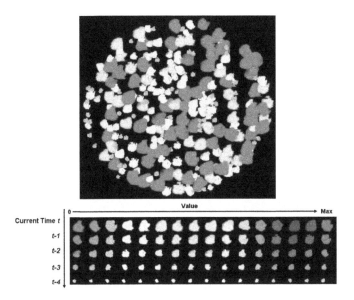

FIG. A.8. Pointillistic visualization showing decrease followed by an increase. The scale allows for the correlation and identification of trends.

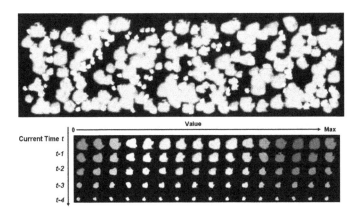

FIG. A.9. Pointillistic visualization showing a constant trend over time. The scale allows for the correlation and identification of trends.

172

REFERENCES

[1] P. K. Agarwal and Kasturi R. Varadarajan. Efficient algorithms for approximating polygonal chains. In *Discrete Comput. Geom., 23*, pages 273–291, 2000.

[2] Maneesh Agrawala and Chris Stolte. Rendering effective route maps: improving usability through generalization. In *SIGGRAPH '01: Proceedings of the 28th annual conference on Computer graphics and interactive techniques*, pages 241–249, New York, NY, USA, 2001. ACM Press.

[3] J. Ahn and K. Wohn. Motion level-of-detail: A simplification method on crowd scene. In *In the Proceedings of CASA 2004*, 2004.

[4] C. L. Bajaj, V. Pascucci, D. Thompson, and X. Y. Zhang. Parallel accelerated isocontouring for out-of-core visualization. In *Proceedings of the 1999 IEEE symposium on Parallel visualization and graphics*, pages 97–104. ACM Press, 1999.

[5] Kenneth P. Bowman. *An Introduction to Programming Using Interactive Data Language (IDL)*. Academic Press, Inc., Orlando, FL, USA, 2005.

[6] S. A. Braun, M. T. Montgomery, and Z. Pu. High-resolution simulation of hurricane bonnie (1998), part i: The organization of eyewall vertical motion. In *J. Atmos. Sci*, volume 63, pages 19–42, 2006.

[7] Stefan Bruckner and Meister Eduard Gröller. Exploded views for volume data. *IEEE Transactions on Visualization and Computer Graphics*, 12(5):1077–1084, 9 2006.

[8] Brian Cabral, Nancy Cam, and Jim Foran. Accelerated volume rendering and tomographic reconstruction using texture mapping hardware. In *Proceedings of the 1994 symposium on Volume visualization*, pages 91–98. ACM Press, 1994.

[9] CDC. Us infant mortality rates by state 1995-2002. US Centers for Disease Control and Prevention, National Center for Health Statistics, 2002.

174

[10] W. S. Chan and F. Chin. Approximation of polygonal curves with minimum number of line segments or minimum error. In *Internat. J. Comput. Geom. Appl.*, pages 59–77, 1996.

[11] Yi-Jen Chiang. Out-of-core isosurface extraction of time-varying fields over irregular fields. In *Proceedings of the conference on Visualization '03*, pages 29–37, 2003.

[12] J. H. Clark. Hierarchical geometric models for visible surface algorithms. In *Communications of the ACM*, pages 547–554, 1976.

[13] Jonathan Cohen, Marc Olano, and Dinesh Manocha. Appearance-preserving simplification. In *SIGGRAPH '98: Proceedings of the 25th annual conference on Computer graphics and interactive techniques*, pages 115–122, New York, NY, USA, 1998. ACM Press.

[14] COLA. Grads, http://www.iges.org/grads/grads.html, 1988.

[15] Carlos Correa, Deborah Silver, and Min Chen. Feature aligned volume manipulation for illustration and visualization. *IEEE Transactions on Visualization and Computer Graphics (Proceedings Visualization / Information Visualization 2006)*, 12(5), September-October 2006.

[16] Balázs Csèbfalvi, Lukas Mroz, Helwig Hauser, Andreas König, and Eduard Gröller. Fast visualization of object contours by non-photorealistic volume rendering. In *Proceedings of the Eurographics Conference '01*, volume 20(3), 2001.

[17] C. A. Davis and L. F. Bosart. Numerical simulations of the genesis of hurricane diana (1984). part i. control simulation. *Monthly Weather Review*, 129(8):1859–1881, 2001.

[18] C. A. Davis and L. F. Bosart. Numerical simulations of the genesis of hurricane diana (1984). part i. sensitivity of track and intensity. *Monthly Weather Review*, 130(5):1100–1124, 2002.

[19] UC Davis. Time varying data repository, 2004.

[20] Helmut Doleisch, Philipp Muigg, and Helwig Hauser. Interactive visual analysis of hurricane isabel with simvis. In *IEEE Visualization 2004 Contest Entry*, 2004.

[21] David Douglas and Thomas Peucker. Algorithms for the reduction of the number of points required to represent a digitized line or its caricature. In *The Canadian Cartographer 10(2)*, pages 112–122, 1973.

[22] P. G. Drazin and W. H. Reid. *Hydrodynamic Stability*. Cambridge University Press, 1981.

[23] David Ellsworth, Ling-Jen Chiang, and Han-Wei Shen. Accelerating time-varying hardware volume rendering using tsp trees and color-based error metrics. In *VVS '00: Proceedings of the 2000 IEEE symposium on Volume visualization*, pages 119–128. ACM Press, 2000.

[24] Victor Fernandez and Deborah Silver. Computational fluid dynamics - turbulent vortex dataset: http://www.caip.rutgers.edu/ xswang/feature/index.html, 1998.

[25] Nathaniel Fout, Kwan-Liu Ma, and James Ahrens. Time-varying, multivariate volume data reduction. In *SAC '05: Proceedings of the 2005 ACM symposium on Applied computing*, pages 1224–1230, New York, NY, USA, 2005. ACM Press.

[26] Thomas A. Funkhouser and Carlo H. Sequin. Adaptive display algorithm for interactive frame rates during visualization of complex virtual environments. In *SIGGRAPH '93: Proceedings of the 20th annual conference on Computer graphics and interactive techniques*, pages 247–254, New York, NY, USA, 1993. ACM Press.

[27] Richard S. Gallagher. *Computer Visualization: Graphics Techniques for Scientific and Engineering Analysis*. CRC Press, Inc., Boca Raton, FL, USA, 1994.

[28] Kenny Gruchalla and Jonathan Marbach. Immersive visualization of the hurricane isabel dataset. In *IEEE Visualization 2004 Contest Entry*, 2004.

[29] Markus Hadwiger, Christoph Berger, and Helwig Hauser. High-quality two-level volume rendering of segmented data sets on consumer graphics hardware. In *Proceedings of the conference on Visualization '03*, pages 40–47. IEEE Computer Society, 2003.

[30] Paul Haeberli. Paint by numbers: abstract image representations. In *Proceedings of the 17th annual conference on Computer graphics and interactive techniques*, pages 207–214, 1990.

[31] Christopher G. Healey, James T. Enns, Laura G. Tateosian, and Mark Remple. Perceptually-based brush strokes for nonphotorealistic visualization. In *ACM Transactions on Graphics*, volume 23 (1), pages 64–96, 2004.

[32] Christopher G. Healey. Formalizing artistic techniques and scientific visualization for painted renditions of complex information spaces. In *In Proceedings International Joint Conference on Artifical Intelligence*, pages 371–376, 2001.

[33] Anders Helgeland and Thomas Elboth. Hurricane visualization using anisotropic diffusion and volume rendering. In *The 5th Annual Gathering on High Performance Computing in Norway - NOTUR*, 2005.

[34] Aaron Hertzmann. Painterly rendering with curved brush strokes of multiple sizes. In *Proceedings of the 25th annual conference on Computer graphics and interactive techniques*, pages 453–460. ACM Press, 1998.

[35] W. Hibbard and D. Santek. The vis5d system for easy interactive visualization. In *Proc. Visualization '90*, pages 28–35, 1990.

[36] Bill Hibbard. VisAD: connecting people to computations and people to people. *Computer Graphics*, 32(3):10–12, August 1998.

[37] Elaine R. S. Hodges. *The Guild Handbook of Scientific Illustration*. John Wiley and Sons, 1989.

[38] Hugues Hoppe. Progressive meshes. In *SIGGRAPH '96: Proceedings of the 23rd annual conference on Computer graphics and interactive techniques*, pages 99–108, New York, NY, USA, 1996. ACM Press.

[39] T.J. Jankun-Kelly and Kwan-Liu Ma. A study of transfer function generation for time-varying volume data. In *Proceedings of Volume Graphics, 2001*, 2001.

[40] Ming Jiang, Naeem Shareef, Caixia Zhang, Roger Crawfis, Raghu Machiraju, and Han-Wei Shen. Visualization fusion: Hurricane isabel dataset. In *IEEE Visualization 2004 Contest Entry*, 2004.

[41] Greg P. Johnson and Christopher A. Burns. Opengl visualization of hurricane isabel. In *IEEE Visualization 2004 Contest Entry*, 2004.

[42] Alark Joshi and Penny Rheingans. Illustration-inspired techniques for visualizing time-varying data. In *Proceedings IEEE Visualization 2005*, pages 679–686, 2005.

[43] Yuya Kawagishi, Kazuhide Hatsuyama, and Kunio Kondo. Cartoon blur: Non-photorealistic motion blur. In *Proceedings of the Computer Graphics International Conference*, pages 276–281. IEEE Computer Society, 2003.

[44] M. Kemp. *The Science of Art*. Yale University Press, 1990.

[45] Gordon Kindlmann, Ross Whitaker, Tolga Tasdizen, and Torsten Moller. Curvature-based transfer functions for direct volume rendering: Methods and applications. In *VIS '03: Proceedings of the 14th IEEE Visualization 2003 (VIS'03)*, pages 513–520, Washington, DC, USA, 2003. IEEE Computer Society.

[46] R. M. Kirby, H. Marmanis, and David H. Laidlaw. Visualizing multivalued data from 2d incompressible flows using concepts from painting. In *VIS '99: Proceedings of the conference on Visualization '99*, pages 333–340, Los Alamitos, CA, USA, 1999. IEEE Computer Society Press.

[47] Robert Kosara, Christopher G. Healey, Victoria Interrante, David H. Laidlaw, and Colin Ware. User studies: Why, how, and when? *IEEE Comput. Graph. Appl.*, 23(4):20–25, 2003.

[48] Bill Kuo, Wei Wang, Cindy Bruyere, Tim Scheitlin, and Don Middleton. Hurricane isabel data, 2004.

[49] David H. Laidlaw, Eric T. Ahrens, David Kremers, Matthew J. Avalos, Russell E. Jacobs, and Carol Readhead. Visualizing diffusion tensor images of the mouse spinal cord. In *VIS '98:*

Proceedings of the conference on Visualization '98, pages 127–134, Los Alamitos, CA, USA, 1998. IEEE Computer Society Press.

[50] Peter Lindstrom and Greg Turk. Image-driven simplification. *ACM Trans. Graph.*, 19(3):204–241, 2000.

[51] Peter Litwinowicz. Processing images and video for an impressionist effect. In *SIGGRAPH '97: Proceedings of the 24th annual conference on Computer graphics and interactive techniques*, pages 407–414, New York, NY, USA, 1997. ACM Press/Addison-Wesley Publishing Co.

[52] Aidong Lu, Christopher J. Morris, David Ebert, Penny Rheingans, and Charles Hansen. Non-photorealistic volume rendering using stippling techniques. In *Proceedings of the conference on Visualization '02*, pages 211–218, 2002.

[53] David P. Luebke and Benjamin Hallen. Perceptually-driven simplification for interactive rendering. In *Proceedings of the 12th Eurographics Workshop on Rendering Techniques*, pages 223–234, London, UK, 2001. Springer-Verlag.

[54] D. Luebke, M. Reddy, J. Cohen, A. Varshney, B. Watson, and R. Huebner. *Level of Detail for 3D Graphics*. Morgan-Kaufmann, Inc., 2003.

[55] Eric B. Lum and Kwan-Liu Ma. Hardware-accelerated parallel non-photorealistic volume rendering. In *Proceedings of the second international symposium on Non-photorealistic animation and rendering*, pages 67–ff, 2002.

[56] Eric B. Lum, Kwan Liu Ma, and John Clyne. Texture hardware assisted rendering of time-varying volume data. In *Proceedings of the conference on Visualization '01*, pages 263–270. IEEE Computer Society, 2001.

[57] Kwan-Liu Ma and David M. Camp. High performance visualization of time-varying volume data over a wide-area network status. In *Supercomputing '00: Proceedings of the 2000*

ACM/IEEE conference on Supercomputing (CDROM), page 29, Washington, DC, USA, 2000. IEEE Computer Society.

[58] Edward MacCurdy. *The Notebooks of Leonardo Da Vinci*, volume 1. The Reprint Society London, 1954.

[59] David. W. Martin. *Doing Psychology Experiments*. Wadsworth Publishinng, 6th edition, 2003.

[60] Scott McCloud. *Understanding Comics*. HarperCollins publishers, 1994.

[61] Barbara J. Meier. Painterly rendering for animation. In *SIGGRAPH '96: Proceedings of the 23rd annual conference on Computer graphics and interactive techniques*, pages 477–484, New York, NY, USA, 1996. ACM Press.

[62] M. T. Montgomery, V. A. Vladimirov, and P. V. Denissenko. An experimental study on hurricane mesovorticies. In *Journal of Fluid Mechanics*, volume 471, pages 1–32, 2002.

[63] Nabil Mustafa, Eleftheris Koutsofios, Shankar Krishnan, and Suresh Venkatasubramanian. Hardware-assisted view-dependent map simplification. In *SCG '01: Proceedings of the seventeenth annual symposium on Computational geometry*, pages 50–59, New York, NY, USA, 2001. ACM Press.

[64] NASA. Global rainfall data (http://data.giss.nasa.gov/), 2000.

[65] NOAA. *Hurricane Basics*. National Hurricane Center, 1999.

[66] Frits H. Post, Frank J. Post, Theo Van Walsum, and Deborah Silver. Iconic techniques for feature visualization. In *Proceedings of the 6th conference on Visualization '95*, page 288. IEEE Computer Society, 1995.

[67] PurpleAmerica. http://www.princeton.edu/ rvdb/JAVA/election2004/, 2004.

[68] Zenon W. Pylyshyn. *Seeing and Visualizing: Its not what you think (Life and Mind)*. Bradford Book, 2003.

180

[69] Penny Rheingans and David Ebert. Volume illustration: Non-photorealistic rendering of volume models. In *IEEE Transactions on Visualization and Computer Graphics*, volume 7(3), pages 253–264, 2001.

[70] RIVA. Riva flv encoder (http://rivavx.de/?encoder), 2006.

[71] A. H. Robinson and R. D. Sale. *Elements of Cartography*. John Wiley and Sons, Inc., 1969.

[72] Bernice Rogowitz and Lloyd Treinish. Data visualization: the end of the rainbow. *Spectrum, IEEE*, 35(12):52–59, 1998.

[73] O. N. Rood. *Modern Chromatics, with Applications to Art and Industry*. Appleton, New York., 1879.

[74] Ravi Samtaney, Deborah Silver, Norman Zabusky, and Jim Cao. Visualizing features and tracking their evolution. *Computer*, 27(7):20–27, 1994.

[75] Natascha Sauber, Holger Theisel, and Hans-Peter Seidel. Multifield-graphs: An approach to visualizing correlations in multifield scalar data. *IEEE Transactions on Visualization and Computer Graphics*, 12(5):917–924, 2006.

[76] P. Coleman Saunders, Victoria Interrante, and Sean C. Garrick. Pointillist and glyph-based visualization of nanoparticles in formation. In *Joint Eurographics/IEEE-VGTC Symposium on Visualization*, pages 169–176, 2005.

[77] T. Schafhitzel, Daniel Weiskopf, and Thomas Ertl. Investigating hurricane isabel using texture advection. In *IEEE Visualization 2004 Contest Entry*, 2004.

[78] M. Schirski, A. Gerndt, T. van Reimersdahl, T. Kuhlen, P. Adomeit, O. Lang, S. Pischinger, and C. Bischof. Vista flowlib - framework for interactive visualization and exploration of unsteady flows in virtual environments. In *Proceedings of the workshop on Virtual environments 2003*, pages 77–85. ACM Press, 2003.

[79] Han-Wei Shen and Christopher R. Johnson. Differential volume rendering: a fast volume visualization technique for flow animation. In *VIS '94: Proceedings of the conference on Visualization '94*, pages 180–187. IEEE Computer Society Press, 1994.

[80] Han-Wei Shen, Ling-Jen Chiang, and Kwan-Liu Ma. A fast volume rendering algorithm for time-varying fields using a time-space partitioning (tsp) tree. In *Proceedings of the conference on Visualization '99*, pages 371–377. IEEE Computer Society Press, 1999.

[81] Deborah Silver and Yashodhan Kusurkar. Visualizing time varying distributed datasets. In *Proceedings of Visualization Development Environment 2000*. IEEE Computer Society Press, 2000.

[82] Deborah Silver and X. Wang. Volume tracking. In *Proceedings of the 7th conference on Visualization '96*, pages 157–ff. IEEE Computer Society Press, 1996.

[83] Deborah Silver and Xin Wang. Tracking and visualizing turbulent 3d features. In *IEEE Transactions on Visualization and Computer Graphics*, volume 3 (2), pages 129–141. IEEE Educational Activities Department, 1997.

[84] Aleksander Stompel, Eric B. Lum, and Kwan-Liu Ma. Feature-enhanced visualization of multidimensional, multivariate volume data using non-photorealistic rendering techniques. In *Proceedings of Pacific Graphics 2002 Conference*, 2002.

[85] M. Strengert, M. Magallón, D. Weiskopf, Stefan Guthe, and T. Ertl. Large volume visualization of compressed time-dependent datasets on gpu clusters. *Parallel Comput.*, 31(2):205–219, 2005.

[86] Philip Sutton and Charles D. Hansen. Isosurface extraction in time-varying fields using a temporal branch-on-need tree (t-bon). In *Proceedings of the conference on Visualization '99*, pages 147–153. IEEE Computer Society Press, 1999.

[87] Nikolai Svakhine, Yun Jang, David Ebert, and Kelly Gaither. Illustration and photography

182

inspired visualization of flows and volumes. In *Proceedings of IEEE Visualization 2005*, pages 687–694, 2005.

[88] L. G. Tateosian, C. G. Healey, and J. T. Enns. Engaging viewers through nonphotorealistic visualizations. In *Fifth International Symposium on Non-Photorealistic Animation and Rendering (NPAR).*, 2007.

[89] S.M.F. Treavett and Min Chen. Pen-and-ink rendering in volume visualization. In *Proceedings of the conference on Visualization '00*, pages 203–210, 2000.

[90] Barbara Tversky, Julie Bauer Morrison, and Mireille Betrancourt. Animation: can it facilitate? *Int. J. Hum.-Comput. Stud.*, 57(4):247–262, 2002.

[91] Barbara Tversky. Distortions in cognitive maps. In *Geoforum, 23(2)*, pages 131–138, 1992.

[92] Tim Urness, Victoria Interrante, Ivan Marusic, Ellen Longmire, and Bharathram Ganapathisubramani. Effectively visualizing multi-valued flow data using color and texture. In *IEEE Visualization 2003*, pages 115–121, 2003.

[93] Ivan Viola, Armin Kanitsar, and Meister Eduard Gröller. Importance-driven feature enhancement in volume visualization. *IEEE Transactions on Visualization and Computer Graphics*, 11(4):408–418, 2005.

[94] Chris Weigle and David C. Banks. Extracting iso-valued features in 4-dimensional scalar fields. In *Proceedings of the Symposium in Volume Visualization*, pages 103–100, 1998.

[95] Rudiger Westermann. Compression domain rendering of time-resolved volume data. In *Proceedings of the 6th conference on Visualization '95*, page 168. IEEE Computer Society, 1995.

[96] Jane Wilhelms and Allen Van Gelder. Octrees for faster isosurface generation. In *Proceedings of the 1990 workshop on Volume Visualization*, pages 57–62. ACM Press, 1990.

[97] Georges Winkenbach and David H. Salesin. Computer-generated pen and ink illustration. In *Proceedings of SIGGRAPH '94*, volume 28 of *Computer Graphics Proceedings, Annual Conference Series*, pages 91–100, 1994.

[98] Jonathan Woodring and Han-Wei Shen. Chronovolumes: a direct rendering technique for visualizing time-varying data. In *Proceedings of the 2003 Eurographics/IEEE TVCG Workshop on Volume graphics*, pages 27–34. ACM Press, 2003.

[99] Jonathan Woodring and Han-Wei Shen. Multi-variate, time varying, and comparative visualization with contextual cues. *IEEE Transactions on Visualization and Computer Graphics*, 12(5):909–916, 2006.

www.ingramcontent.com/pod-product-compliance
Lightning Source LLC
Chambersburg PA
CBHW071423050326

40689CB00010B/1950